PLASMA SPRAYING

Theory and Applications

PLASMA SPRAYING

Theory and Applications

Editor

R Suryanarayanan

Laboratoire de Physique des Solides, de Bellevue, CNRS

World Scientific

Singapore • New Jersey • London • Hong Kong

Published by

World Scientific Publishing Co. Pte. Ltd.
P O Box 128, Farrer Road, Singapore 9128
USA office: Suite 1B, 1060 Main Street, River Edge, NJ 07661
UK office: 73 Lynton Mead, Totteridge, London N20 8DH

PLASMA SPRAYING: THEORY AND APPLICATIONS

ISBN 981-02-1363-8

Printed in Singapore

CONTENTS

PREFACE

Plasma spraying, which grew out of oxyacetylene flame spray, not only provides coatings that enhance the service life of critical components exposed to an aggressive atmosphere but also can produce near-net shapes of certain complex materials. For several years, the success of plasma spraying mainly depended on the "skill" of certain experienced operators. Recently, however, a number of groups have been conducting fundamental research to understand different parameters that affect the quality of the coating. Some of these are heat and momentum transfer of flame particles, plasma power, plasma gas, grain structure and interface bonding.

Modern physical methods, computing techniques and automation have been used to better control and understand what goes on between the particle injection and deposit formation and quality. This has increased the reproducibility and reliability of the coating properties and has extended the applications. The recent advances in computation and optimization of the spray process are dealt with in great detail in Chapters 1 to 5. Though most of the present applications are directed towards aeronautic, mechanical and chemical industries, there has recently been interest in using plasma spraying for electronic materials. These, however, impose stringent conditions on the sprayer and the sprayed material. Three examples are given in this volume to illustrate possible future applications: silicon for the semiconductor industry, such as in solar cells; Sm-Co for permanent magnets; and the newly discovered high critical temperature superconductors, for which $YBa_2Cu_3O_7$ is a typical promising material for magnetic shielding and other applications (Chapters 9 to 12).

Perhaps several changes in the design of the plasma torch and the powder distributor are necessary in order to eliminate some impurities (e.g. Al, Ni, Fe), especially in the case of silicon coatings. Further post deposition techniques, such as electron beam glazing and halogen lamp annealing, are being developed to increase the grain size, improve the microstructure, and in some cases eliminate certain impurities, which will definitely enhance the potential of plasma spraying and make it attractive for material processing industries concerned with these applications.

This book also examines an alternative technique for plasma spraying — high velocity oxy-fuel spraying, which seems to be quite successful in certain cases (Chapter 8). Chapters 6 and 7 examine in detail the structure property of ceramic coatings, especially in the case of copper electroplated alumina coating.

Since the contributing authors are from different parts of the world, one will catch a glimpse of research going on in the field in their respective countries.

In conclusion: this volume covers the fundamentals of plasma spraying and a wide range of applications, from metallurgical to biomedical to electronic engineering. Thus I sincerely hope that the book will be a useful reference and guide for advanced graduate students in materials science and engineering (including electronic engineering), research scientists and engineers in universities, research establishments and various industries.

I would like to thank all the authors for expending their time and effort in preparing their valuable contributions. In addition I would like to thank Prof K. K. Phua, Ms Faridah Shahab and Mr H. T. Leong for encouraging me, and I appreciate very much the patience shown and the help given during the editing of this volume.

R. Suryanarayanan
December 1992

Chapter 1

FUNDAMENTALS OF PLASMA PARTICLE MOMENTUM AND HEAT TRANSFER

MAHER I. BOULOS
Department of Chemical Engineering, University of Sherbrooke
Sherbrooke, Quebec, CANADA

PIERRE FAUCHAIS AND ARMELLE VARDELLE
Faculté des Sciences, Université de Limoges
Limoges, FRANCE

and

EMIL PFENDER
Department of Mechanical Engineering
University of Minnesota
Minneapolis, Minnesota, USA

ABSTRACT

Plasma-particle interactions represent one of the most important aspects on which the outcome of most plasma processing of materials depends. It plays a key role in such applications as in-flight melting, densification and spheroidization of powders, atmospheric and vacuum plasma spraying of protective coatings, plasma deposition of near net shape bodies, d.c. and r.f. induction plasma deposition of metal matrix composites, and plasma reactive deposition. A review is presented of recent developments in our understanding of the basic momentum and heat transfer phenomena involved and the way they can influence the overall outcome of the process. Both experimental and theoretical studies are presented in four separate sections; particle injection, plasma-particle momentum and heat transfer for a single particle in a plasma flow, single particle trajectories and temperature histories, and plasma-particle interactions under dense loading conditions. The discussion covers such phenomena as the effect of the particle carrier gas on the flow and temperature fields in the plasma reactor, the effect of steep temperature gradients on plasma-particle momentum and heat transfer, internal heat conduction within solid and porous particles, particle evaporation, and non-continuum effects.

1. Introduction

In the thermal processing of powders under plasma conditions whether for spheroidization, melting and densification or the plasma spraying of protective coatings or free standing bodies, the proper control of the trajectories of the particles and of their temperature histories in the plasma represents one of the crucial aspects on which the overall success of the operation strongly depends. In fact a slight deviation from near optimal conditions can easily lead to poor results due to either the lack of melting of the powder, or the modification of its composition due to partial evaporation.

This chapter is mainly devoted to a review of the different fundamental phenomena encountered in the plasma processing of powders. This includes a discussion of particle injection techniques followed by an analysis of plasma-particle momentum and heat transfer. Diffusion phenomena taking place in the vapor cloud surrounding a single particle is discussed with the coupling of the heat and mass transfer equations during the transient heating of the particles.

Single particle trajectory calculations in d.c. plasma jets and r.f. inductively coupled discharges is discussed next followed by an analysis of plasma-particle interactions under dense loading conditions.

2. Particle injection

A number of techniques has been used for particle injection in a plasma jet. These techniques varied mostly in the location of the point of injection and the angle of attack with respect to the axis of the jet. A schematic diagram of a typical plasma spraying system, is given in Fig 1. In this case the powder is injected in the anode constriction region with a slight inclination towards the upstream side of the jet. The powder is transported with a carrier gas through injectors which typically have an i.d. between 1.6 and 2.4 mm.

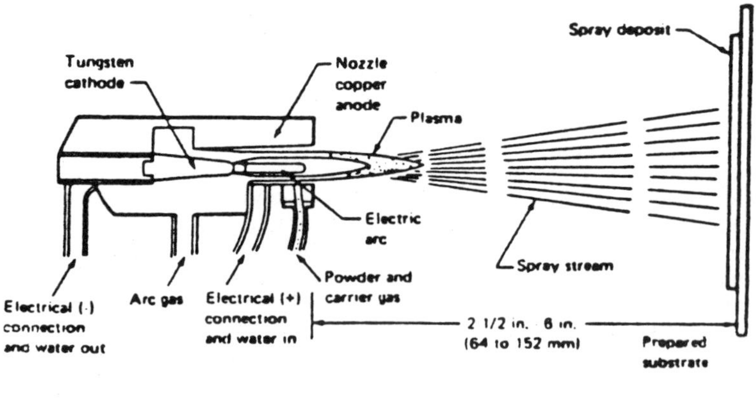

Figure 1: Schematic sketch of a plasma spray torch.

Other powder injection configurations have also been used in which the powder is injected at right angles to the plasma jet at the downstream end of the anode. A comparison of the powder injector either orghogonal to the plasma jet or tilted at 60° against the flow is given in Fig. 2.a. To obtain similar trajectories with the two injectors (of the same internal diameter : 1.6 mm) the argon powder carrier gas flowrate $\dot{m}_{cg}$ was 3.5 slm for the orthogonal injector and only 2 slm with the tilted one, otherwise the mean trajectory (defined later on) crosses the plasma jet too rapidly. This tilting has a drastic influence on the particles' mean velocity and temperature history along their mean trajectory as illustrated in Fig. 2.b and c. In spite of a higher velocity with the tilted injector, the particles reach a higher surface temperature due to their longer residence time in the hottest zone of the jet. However, due to the lower injection velocity with the counter current injector and injection in a hotter zone of the plasma (compared to the orthogonal injection), sticking problems arise, reducing the interest of this type of injection.

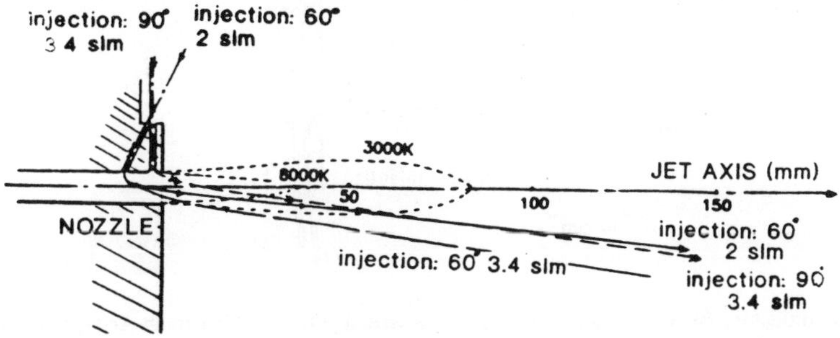

Figure 2.a : Mean trajectories for calcia stabilized zirconia particles (+44-75μm) injected at two angles in an Ar-H_2 plasma jet, (75 slm Ar, 15 slm H_2, I=400 A, V=72 V, ρ=60%, injector i.d. 1.6 mm, nozzle i.d. 8 mm, after Vardelle et al [2].

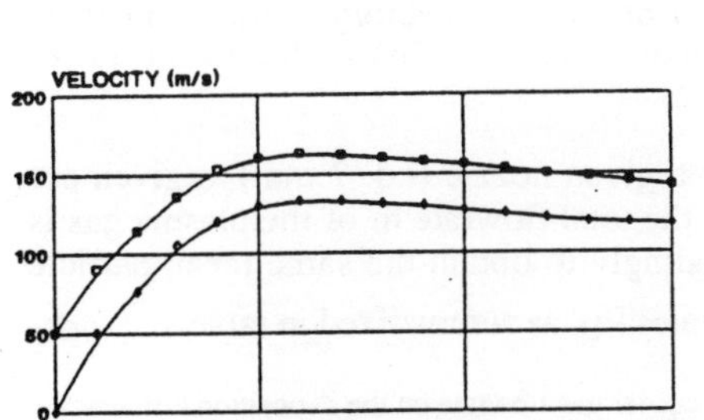

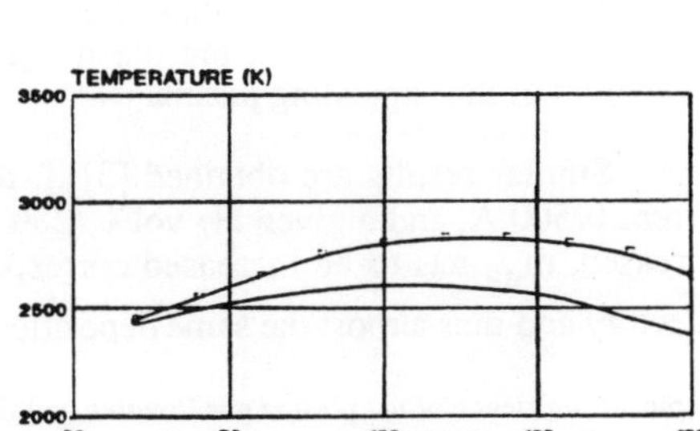

Figure 2.b and c : Mean velocity (left) and surface temperature (right) along the mean trajectories for the two injection angles, after Vardelle et al [2].

In either case, special attention has to be given to the proper control of the injection velocity of the particles in the plasma jet in order to insure that they penetrate the hot region of the plasma and acquire as close as possible an axial trajectory.

For a given plasma jet (nozzle diameter, power level, gas nature and flowrate) the momentum of the injected particles has to be adapted to that of the plasma jet in order to insure that their residence time is longest in the hot zones of the jet. This is illustrated in Fig. 3 where, for a given plasma jet, a given carrier gas (Ar in the following), a given injector and a given fused and crushed (FC) alumina powder size distribution, very different mean trajectories can be observed depending on the carrier gas flow rate $\dot{m}_{cg}$. When $\dot{m}_{cg}$ is too low, the particles do not penetrate into the jet and when it is too high they cross it.

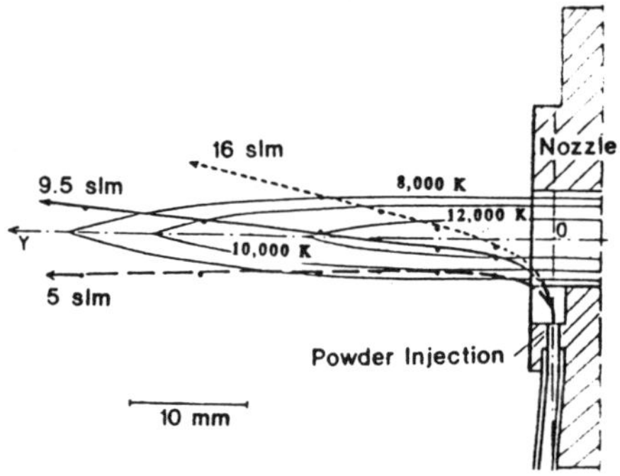

Figure 3: Mean trajectory for Yttria stabilized zirconia (8 wt% Y_2O_3) particles' in an Ar-H_2 plasma jets, (75 slm Ar, 15 slm H_2, nozzle i.d. 8 mm, I=450 A, V=71, ρ=60%, FC powder +8-25 μm, injector i.d. 1.6 mm), after Gyenis et al [3].

To illustrate the influence of the plasma jet momentum, the trajectories of particles with the injector either in the nozzle, 3 mm upstream of its exit (internal injection) or outside 5 mm downstream of its exit (external injection), are given in Fig. 4. With the same carrier gas flowrate as that allowing an optimum trajectory with the internal injection, the particles cross the plasma jet more rapidly. This is due to the lower momentum of the expanding plasma jet.

Similar results are obtained [3] if, for a given nozzle (i.d. 7 mm), a given arc current I=500 A, and a given H_2 vol% (25%), the total flowrate $\dot{m}$ of the plasma gas is increased, $\dot{m}_{cg}$ has to be increased correspondingly to obtain the same mean particle trajectory and thus almost the same deposition rates ρ_D, as summarized in table 1.

Table 1 : Influence of the plasma gas flowrate and the carrier gas flowrate on the deposition efficiency.

P (kW)	35	35	35
$\dot{m}$ (slm)	46.7	53.3	60
$\dot{m}_{cg}$ (slm)	4.3	5.5	7
ρ_D (%)	56	57	58

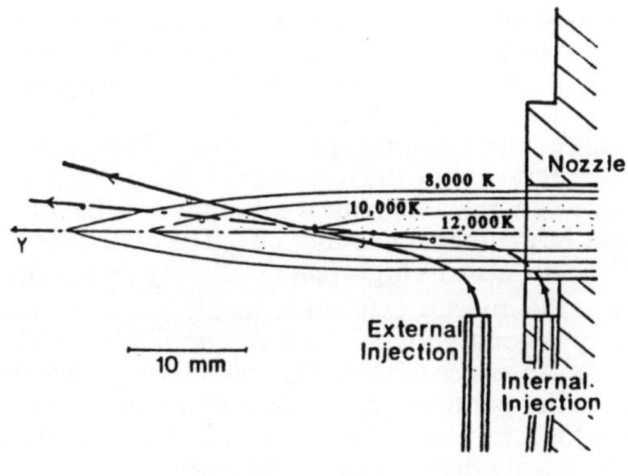

Figure 4: Mean trajectory of calcia stabilized zirconia (7.5 wt%CaO, FC -15+5 μm) particles' injected, in a d.c. plasma jet under the same conditions as defined in Fig. 3.

For a given powder and a given injector geometry, the smaller the particles, the higher must be the carrier gas flowrate. This is illustrated in Fig. 5 a and b where for two size ranges (-21+15 μm and - 90+45 μm) of FC alumina particles injected with two carrier gas flowrates, the measured velocity distributions at the injector exit gives the same mean trajectory. It should be noted that the velocities given in Fig. 5 were measured by LDA in the direction of the injector axis.

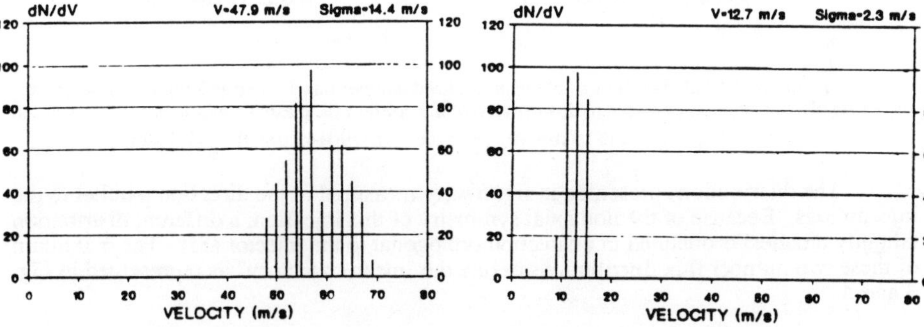

Figure. 5: Injection velocity distributions for FC alumina particles which gives rise to the same mean trajectory in an Ar-H_2 plasma jet : 75 slm Ar, 15 slm H_2, nozzle i.d. 8 mm, I-400 A, V=72 V, ρ=60% injector i.d. 1.6 mm. (a) -21+15 μm, $\dot{m}_{cg}$=7.5 slm, (b) -90+45 μm, $\dot{m}_{cg}$ = 2.5 slm [4].

The pneumatic transport of powders in such small diameter tubes on the other hand, suffers from the excessive loss of particle kinetic energy through particle-wall interactions which results in a large slip between the gas and the particle transport velocities. For a 2 mm i.d. tube the average particle velocity in the transport line can be as low as one third that of the transport gas velocity. The value of the slip velocity, however, drops with the increase of the diameter of the powder transport line and the average particle velocity can reach its typical asymptotic value of 0.6 to 0.8 that of the transport gas with tubes larger than 6 to 8 mm i.d.. The collisions between the particles and the injector wall result also in high radial velocity components and the particle dispersion cone angle at the injector exit can be as high as 30°. The resulting particles trajectories are finally strongly dependent on both size and injection velocity distributions. It is therefore not surprising that even with powders with a very narrow size distribution, the smaller particles, due to their broad injection velocity distribution, exhibit a much broader trajectory distribution (see Fig. 6.a) than the larger particles (see Fig. 6.b) for which the velocity distribution is narrower.

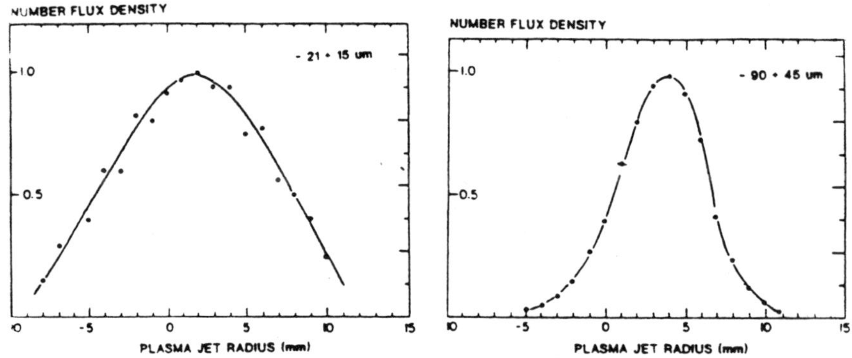

Figure 6: Radial distributions of the normalized number flux for two FC alumina powders in an Ar-H_2 plasma jet measured 75 mm downstream of the nozzle exit, plasma conditions are the same as given in Fig. 5.(a) -21+15 μm, $\dot{m}_{cg}$=7.5 slm ; (b) -90+45 μm, $\dot{m}_{cg}$=2.5 slm.

The distributions presented in fig. 6 were measured in the direction parallel to the injector axis. Because of the non axial symmetry of this injection, a different distribution (slightly broader) is obtained in a direction orthogonal to the injector axis. The maximum of these two number flux distributions define the "mean trajectory" as represented in Fig. 3 and 4.

Another important factor, which can have a substantial influence on the efficiency of the powder heating and melting in the plasma, is the effect of the powder carrier gas on the flow and temperature fields in the plasma. The higher $\dot{m}_{cg}$ and the smaller the i.d. of the injector, the higher will be the plasma perturbation. For example an Ar-H_2 plasma jet (45 slm Ar, 15 slm H_2, 35 kW, nozzle i.d. 7 mm) starts to loose its symmetry for an argon carrier gas flowrate higher than 2.5 slm with an injector i.d. of 1.8 mm [5].

2D laser imaging [6], providing information on the particle distribution within the plasma jet, have shown the advantage, especially for small particles (high $\dot{m}_{cg}$), of the use of two powder injectors opposing to each other. In this case, provided two powder feeders are used, the powder distribution within the plasma jet exhibits a cylindrical symmetry and the deposition efficiency is slightly raised (by about 5%).

Problems may also arise when dealing with a powder composed of a heterogeneous mixture of two powders of different particle densities. In this case considerable particle segregation can take place, resulting in the deposition of the two powders on different locations on the substrate as shown on the top two sketches of Figure 7. One remedy that could be used is multi-port injection in which the individual powders are injected at different locations around the periphery of the anode whether equally distributed or clustered on one side of the torch as shown, respectively, on the two lower sketches of Figure 7. In the latter case, the injection velocity of each powder has to be individually controlled in such a way that the trajectory of the particles coincide prior to their impaction on the substrate thus forming a uniform deposit [7].

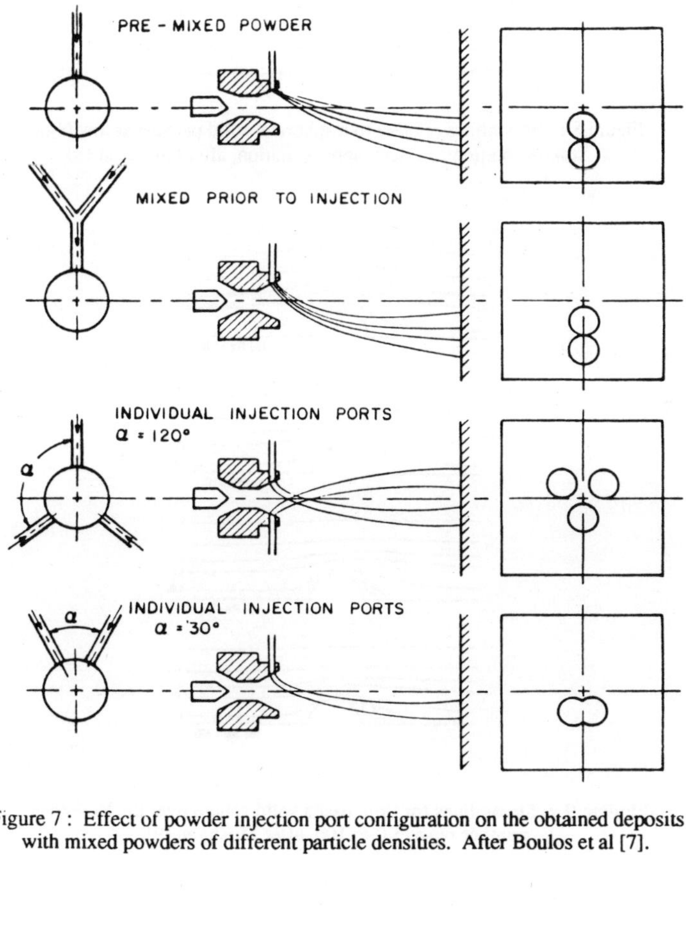

Figure 7 : Effect of powder injection port configuration on the obtained deposits with mixed powders of different particle densities. After Boulos et al [7].

3. Plasma particle momentum and heat transfer

3.1 Introduction

In this section we will deal mainly with the problem of momentum and heat transfer between a plasma and a single particle. For simplicity the analysis will be limited to spherical particles. Few data are available in the literature [8] for particles of other shapes. Attention is also given to internal heat conduction in the particle as well as particle melting and evaporation under plasma conditions.

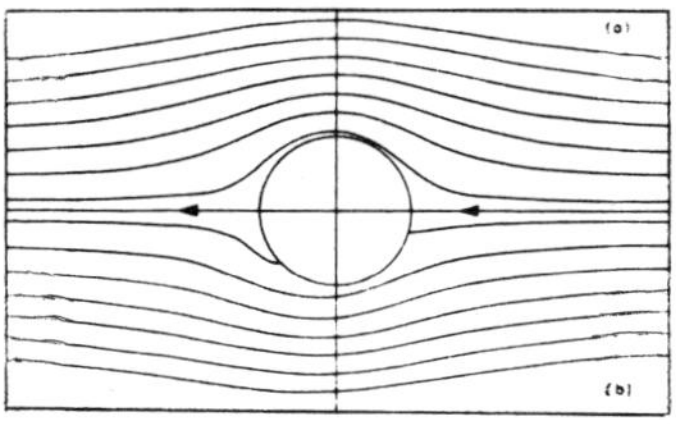

Figure 8 : Streamlines relative to a spherical solid particle at low N_{Re}.
a) Stoke's solution b) Oseen approximation, after Clift et al (8).

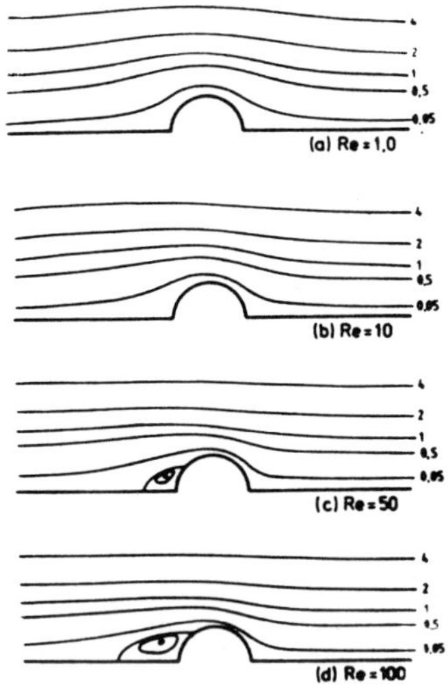

Figure 9 : Streamlines for flow past a solid sphere over the Reynolds number range 1.0 to 100, after Mosliyah [9].

3.2 Flow and temperature fields around a single sphere

Considerable attention has been given to the study of the flow and the temperature fields around single spheres [8]. Typical flow patterns as a function of the particle Reynolds number, N_{Re}, are given in Figures 8 & 9. These are limited to values of $N_{Re}<100$ which are normally encountered under plasma conditions where:

$$N_{Re} = \frac{\rho U_R d_p}{\mu} \tag{1}$$

ρ fluid density
μ fluid viscosity
d_p particle diameter
U_R relative velocity between the particle and its surrounding:

$$U_R = \sqrt{(U-U_p)^2 + (V-V_p)^2}$$

U_p & U particle and gas velocity components in the axial direction, respectively
V_p & V particle and gas velocity components in the radial direction, respectively

It may be noted from Fig. 8 that the increase of the particle Reynolds number, is accompanied by a gradual shift of the flow pattern from the typical symmetric Stokes flow regime shown in Fig. 8 a for $N_{Re}<0.01$, to the viscous but slightly non-symmetric (upstream vs downstream) flow pattern represented by the Oseen approximation for Reynolds number between 0.01 and 1.0, shown in Fig. 8.b. At Reynolds numbers beyond 1.0, Fig. 9, the flow pattern, while still being in the laminar region, becomes increasingly non-symmetric with the eventual development of a separation bubble behind the particle at Reynolds number higher than 50.

Obviously these changes in the flow pattern around the particle will have a direct influence on the drag coefficient , C_D, and the heat transfer coefficients, h, between the particle and its surrounding, where:

$$C_D = \frac{F_D/A}{\frac{1}{2}\rho U_R^2} \tag{2}$$

with:

F_D the total drag force exerted by the fluid on the particle

A the particle projected surface area perpendicular to the flow $(A = \pi d_p^2/4)$

Typical results for C_D as function of the Reynolds number N_{Re} are given in Fig. 10. These can be approximated for Reynolds numbers less than 200 by the correlation given in Table 2.

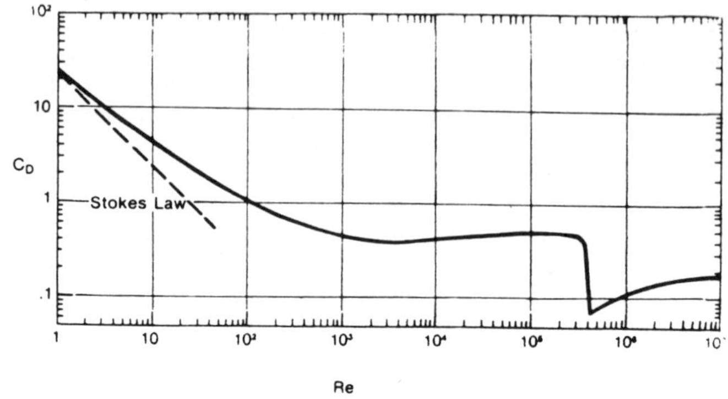

Figure 10: Drag coefficient as function of the Reynolds number for a single sphere, after Clift et al [8].

Table 2: Drag coefficient as function of the Reynolds number for a single sphere

$C_D = \frac{24}{N_{Re}}$	$N_{Re} < 0.2$
$C_D = \frac{24}{N_{Re}} [1 + 0.187\, N_{Re}]$	$0.2 \leq N_{Re} < 2.0$
$C_D = \frac{24}{N_{Re}} [1 + 0.110\, N_{Re}^{0.81}]$	$0.2 \leq N_{Re} < 20.0$
$C_D = \frac{24}{N_{Re}} [1 + 0.189\, N_{Re}^{0.62}]$	$20.0 \leq N_{Re} < 200.0$
$C_D = \frac{24}{N_{Re}} + \frac{6}{1 + \sqrt{N_{Re}}} + 0.4$	$N_{Re} \geq 200$

The heat transfer coefficient between the plasma and the particles is expressed in terms of the Nusselt number defined as:

$$Nu = \frac{h\, d_p}{\kappa} \tag{3}$$

with:

h heat transfer coefficient

κ thermal conductivity of the fluid

Typical data of the Nusselt as function of the Reynolds number at low temperature difference are given in Fig. 11. These can be correlated by the following relationship for Reynolds numbers lower than 200.

$$Nu = 2.0 + 0.6\, N_{Re}^{0.5}\, N_{Pr}^{0.33} \qquad N_{Re} < 200, \text{ and } 0.5 < N_{Pr} < 1.0 \tag{4}$$

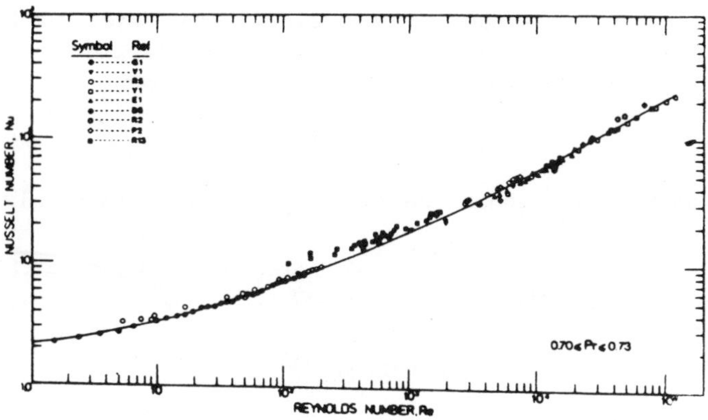

Figure 11: Nusselt number as function of the Reynolds number for a single sphere in air, after Clift et al [8].

3.3 Momentum and heat transfer under plasma conditions

3.3.1 Influence of the temperature gradients

The principal characteristics of momentum and heat transfer under plasma conditions are the presence of charged particles and of steep temperature gradients across the boundary layer which can have a substantial effect on the flow and temperature fields around the sphere through its influence on the local fluid properties.

a) Momentum transfer

The gradient effect is demonstrated in Fig. 12 after Sayegh and Gauvin [11] who computed the flow and temperature fields for a particle Reynolds number of 50, assuming in one case constant fluid properties and taking into account in the second case property variations across the boundary layer (lower and upper parts of the figures, respectively). The parameter T_o was defined as the ratio of the temperature of the sphere surface, T_s, to that of the gas stream T_∞.

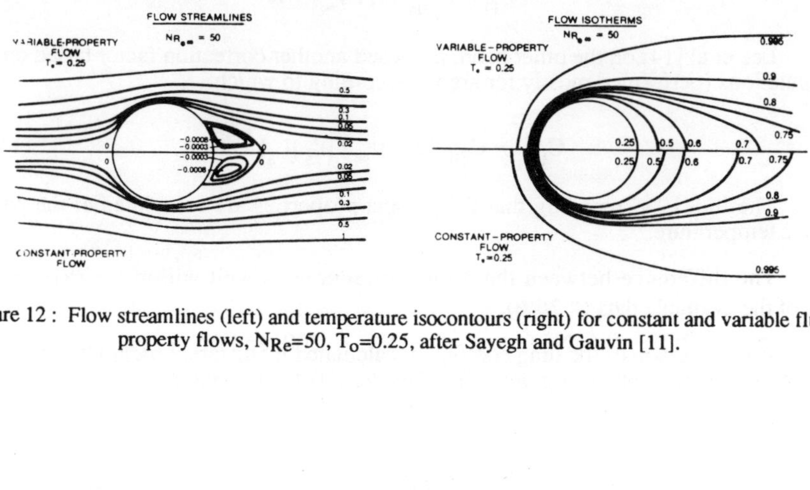

Figure 12 : Flow streamlines (left) and temperature isocontours (right) for constant and variable fluid property flows, N_{Re}=50, T_o=0.25, after Sayegh and Gauvin [11].

Based on their results (obtained for an argon plasma), Sayegh and Gauvin [11] recommended the estimation of the fluid properties (kinematic viscosity, ν, and thermal conductivity, κ) at a reference temperature defined as, $T_{0.19}$, where:

$$T_{0.19} = T_s + 0.19\,(T_\infty - T_s) \tag{5}$$

They modified the standard Nusselt number correlation given in Table 2 to fit their data as follows:

$$Nu_o = 2.0\, f_o + 0.437\, N_{Re}^{0.552}\, N_{Pr}^{m} \tag{6}$$

with

$$m = 0.78\ N_{Re}^{0.552}\ N_{Pr}^{m} \tag{7}$$

and

$$f_o = \left(1 - T_o^{1+x}\right) / \left[(1+x)\left(1 - T_o\right) T_o^{x}\right] \tag{8}$$

Where x is the value of the exponent on T in the expression relating the viscosity and the thermal conductivity to the absolute temperature ($x = 0.8$ for Ar at $T < 10\,000$ K and $p = 10^5$ Pa).

Considerably simpler corrections of the drag coefficient and the heat transfer coefficient have been widely accepted in the literature [12].

One such correction factor proposed by Lewis and Gauvin [13] involves the estimation of the drag coefficient based on the arithmetic mean film temperature across the boundary layer.

$$T_f = \left(T_s + T_\infty\right) / 2.0 \tag{9}$$

and correcting the obtained drag coefficient using the ratio of kinematic viscosity estimated at the mean film temperature, ν_f, to that estimated at the free stream temperature, ν_∞ to the exponent 0.15; i.e.:

$$C_D = C_{DF}\ (\nu_f / \nu_\infty)^{0.15} \tag{10}$$

Lee et al [14] on the other hand, proposed another correction factor based on their computations (performed mostly for argon) according to which:

$$C_D = C_{Df}\left(\rho_\infty\, \mu_\infty / \rho_s\, \mu_s\right)^{0.45} \tag{11}$$

The index "s" meaning that the plasma properties are calculated at the particle surface temperature.

The difference between these two corrections is well within the experimental error of the available data (± 25%).

A comparison of the drag coefficient calculated by different methods for an argon plasma flow over a sphere is given, after Pfender [10,15] in table 3.

Table 3: Drag coefficient for an argon plasma calculated using different correction factors.

T_w [K]	T_∞ [K]	N_{Re} =	0.1	1.0	10.0	20.0	50.0
1000	4000	Simulation	100.8	15.4	2.9	1.8	1.0
		Lee [16] Film temp.	100.6	12.3	2.4	1.6	1.1
		Lee et al. [14]	115.6	14.2	2.7	1.9	1.2
		Lewis & Gauvin [13]	146.1	17.9	3.4	2.4	1.6
2500	10000	Simulation	151.9	17.6	3.3	2.1	1.2
		Film temp.	112.9	13.7	2.5	1.7	1.1
		Lee et al.	136.9	16.6	3.1	2.1	1.4
		Lewis & Gauvin	164.5	19.9	3.7	2.5	1.7
3000	12000	Simulation	199.9	22.4	4.2	2.6	1.4
		Film temp.	137.6	16.2	2.9	1.9	1.2
		Lee et al.	202.3	23.9	4.2	2.8	1.8
		Lewis & Gauvin	226.8	26.8	4.7	3.2	2.0

b) Heat transfer

The particle temperature is governed by a balance between conduction and convection heat transfer between the plasma and the particle less radiative heat losses from the particle to the surrounding. In most cases the radiative transfer from the plasma gas can be neglected because the plasma is optically thin [17]

$$Q = h\left(\pi d_p^2\right)\left(T_\infty - T_s\right) - \left(\pi d_p^2\right) \epsilon \sigma_s \left(T_s^4 - T_a^4\right) \quad (12)$$

where T_s and d_p are respectively the surface temperature and diameter of the particle,

ϵ is the particle emissivity
σ_s the Stephan-Boltzmann constant
T_a the temperature of the surrounding

Assuming a uniform particle temperature implies that the thermal conductivity of the particle material κ_p is much higher than that of the gas κ, i.e.

$$B_i = \frac{\kappa}{\kappa_p} < 0.01 \quad (13)$$

Because of the high temperature gradients and the strong non linear variations of κ with temperature [18] one of the key problems for the heat transfer coefficient calculation through the Nusselt number (eq. (3)) is to determine at which temperature it has to be calculated: film temperature (eq. (9), $T_{0.19}$ (eq. (5)) or any other temperature. This problem is especially crucial when considering diatomic gases for which a strong peak of κ shows up at dissociation temperatures.

A detailed theoretical analysis of the problem of pure conduction by Bourdin et al [19] showed that the equation given in Table 2 holds even under plasma conditions provided that the properties are evaluated as an integrated mean value defined as:

$$\overline{\kappa} = \frac{I}{\left(T_\infty - T_s\right)} \int_{T_s}^{T_\infty} \kappa(T)\, dT \tag{14}$$

It is interesting to note that if the thermal conductivity, κ is a linear function of temperature, Eq. 14 reduces to the commonly used practice of evaluating the property values at the arithmetic mean i.e. $\overline{\kappa}(T) = \kappa(T_f)$, or film temperature.

The application of Eq. (14) can be considerably simplified by splitting the integral with respect to some reference temperature, $T_o = 300$ K, as follows:

$$\overline{\kappa} = \frac{I}{\left(T_\infty - T_s\right)} \left[\int_{T_{300}}^{T_\infty} \kappa(T)\, dT - \int_{T_{300}}^{T_\infty} \kappa(T)\, dT \right] \tag{15}$$

$$\overline{\kappa} = \frac{I}{\left(T_\infty - T_s\right)} \left[I\left(T_\infty\right) - I\left(T_s\right) \right] \tag{16}$$

Values of I(T), known as the heat conduction potential, are given in Fig. 13 for different gases at atmospheric pressure as a function of temperature.

It is worth to notice that the analysis of Bourdin et al [19] is restricted to pure conduction, and in most practical cases for plasma spraying ($N_{Re} < 10$ and $N_{Pr} < 1$) the convection term represents less than 25% of the heat transfer [17].

To account for the temperature gradients Sayegh and Gauvin [11] proposed to calculate the fluid properties at the reference temperature $T_{0.19}$ defined in eq. [5] and to use:

$$Nu = 2.0\ f_o + 0.473\ N_{Pr}^{\,m}\ N_{Re}^{\,0.552} \tag{17}$$

where m and f_o are given by equations (7) and (8)

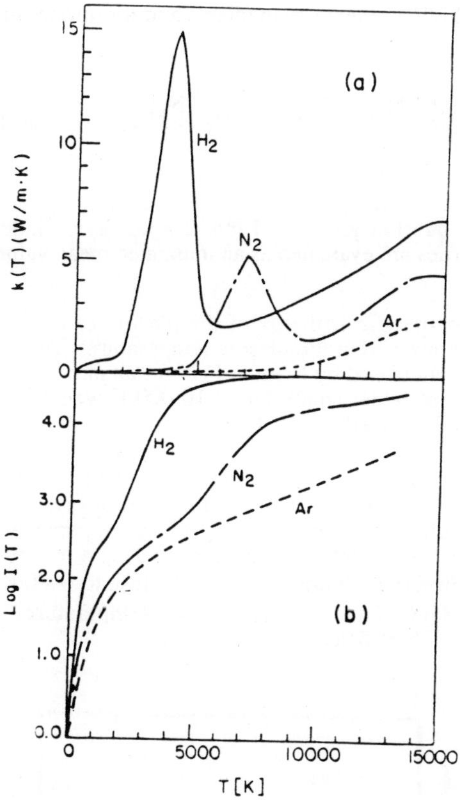

Figure 13: Thermal conductivity (a) and the heat conduction potential (b) for different gases at atmospheric pressure as function of temperature, after Bourdin et al [19].

Lewis and Gauvin [13] suggested for pure argon,

$$Nu = \left(2 + 0.515\ N_{Re}^{1/2}\right)\left(\nu_f/\nu_\infty\right)^{0.15} \tag{18}$$

where T_f is the film temperature at which N_{Re} is calculated. Fizsdon [20] suggested to use

$$Nu = \left(2 + 0.6\ N_{Re_f}^{0.5}\ N_{Pr}^{1/3}\right)\left(\frac{\rho_\infty \mu_\infty}{\rho_s\ \mu_s}\right)^{0.6} \tag{19}$$

where N_{Re} and N_{Pr} are calculated at the film temperature. Lee [14] proposed to use

$$Nu = \left(2 + 0.6\ N_{Re}^{1/2}\ N_{Pr}^{1/3}\right)\left(\frac{\rho_\infty\ \mu_\infty}{\rho_s\ \mu_s}\right)^{0.6}\left(\frac{c_{p\infty}}{c_{ps}}\right) \tag{20}$$

and finally Chen et al [21-25] gave an even more complex relationship.

$$\mathrm{Nu} = 2\left[1 + 0.63\, \mathrm{N}_{\mathrm{Re}_\infty}\, \mathrm{N}_{\mathrm{Pr}_\infty}^{0.8} \left(\mathrm{N}_{\mathrm{Pr}_s} / \mathrm{N}_{\mathrm{Pr}_\infty}\right)^{0.42} \left(\frac{\rho_\infty\, \mu_\infty}{\rho_s\, \mu_s}\right)^{0.52} C^2\right] \tag{21}$$

in which

$$C = \left[1 - \left(h_s / h_\infty\right)^{1.1}\right] \Big/ \left[1 - \left(h_s / h_\infty\right)^{2}\right] \tag{22}$$

Where h is the specific enthalpy of the plasma calculated respectively at the particle surface temperature and at the plasma temperature. The values of the different resulting Nu have been calculated by Pfender [12] for an atmospheric argon plasma. At 4000 K the differences are rather small, but at 10 000 K where ionization is important, they are more drastic as illustrated in Fig. 14.

These differences are particularly important when considering diatomic gases such an A_r - H_2 (17 vol %) mixture. This is illustrated in Fig. 15 representing the heat transfer coefficient calculated by Vardelle [17] versus temperature for a 50 μm particle, with T_S = 1 000 K, travelling at 50 m/s. Correlation 1, 2, 4 in figure 15 were evaluated at T_f and they exhibit similar evolution with a peak close to 6000 K corresponding to T_f = (6000 + 1000) /2 = 3500 K i.e. the dissociation temperature of hydrogen where κ exhibits as well as c_p a high peak as shown in Fig. 13 a.

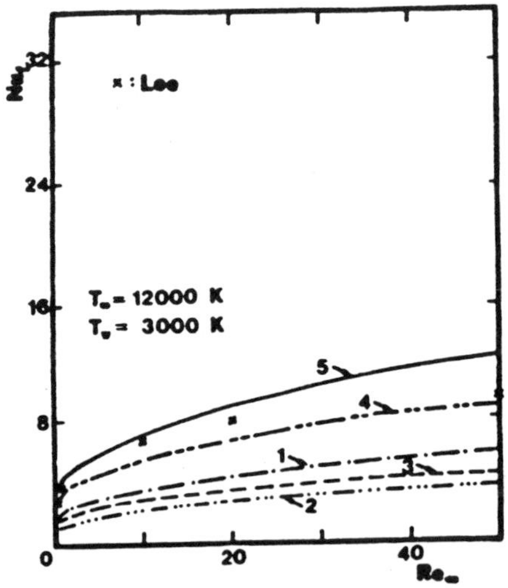

Figure 14: Nusselt number derived by different authors and by computer simulation [12]. 1 - Lewis and Gauvin [13], 2 - Fizsdon [20], 3 - Sayegh and Gauvin [11], 4 - Lee et al [14], 5 - Bourdin et al [19]

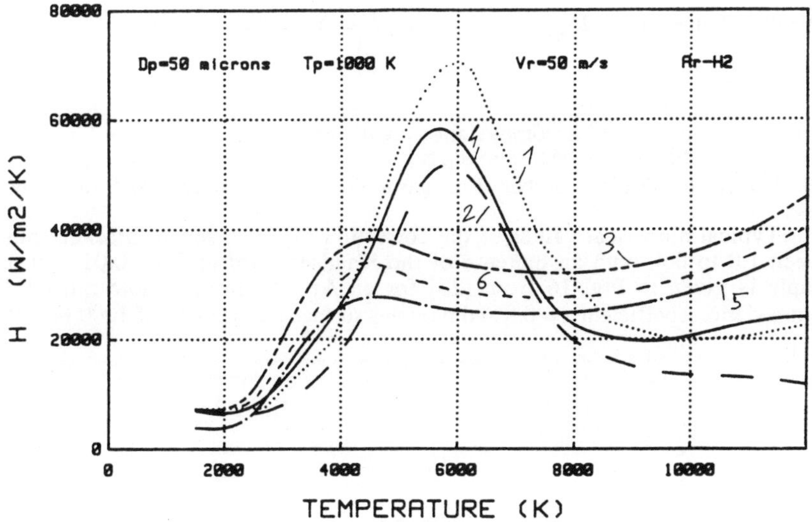

Figure 15: Heat transfer coefficient for a 50 μm diam. particle travelling at 50 m/s with T_s = 1000 K in a A_r-H_2 (17 vol %) plasma (1) Lewis and Gauvin [13], (2) - Fizsdon [20], (3) Convection and induction calculated with integrated properties, (4) Lee et al [14], (5) Nu = 2 : Bourdin et al [19], (6) Chen [21].

It can be seen that the difference between the values of h neglecting convection (curve 5 in Fig. 15) and that taking it into account (curve 3 in Fig. 15) is less than 20% -.

Finally the question is what relationship has to be used . Chen [22] has tried to answer this question by comparing the compiled results from the 8 different relationships he found in the literature to experimental results for spheres respectively 6 mm in diameter in an argon plasma jet, and 8 mm in an air plasma jet. The results, however, in a limited temperature range (7000 to 9200 K) due to the experimental conditions, show that the best fit is obtained with eq. [21].

3.3.2 Influence of the non continuum effects

Other corrections which might be necessary when dealing with the transport and heating of very fine powders (d_p<10 μm) under plasma conditions whether at atmospheric pressure or soft vacuum conditions, are due to non-continuum effects. These can be particularly important when the mean free path of the plasma, λ, is of the same order of magnitude as the diameter of the particles, d_p. According to Chen and Pfender [23] the Knudsen effect should be taken into account in the Knudsen number regime $0.01 < Kn < 1.0$ where, $Kn = \lambda / d_p$.

The following correction for the drag coefficient was proposed [24]

$$C_D = (C_D)_{cont.} \left[\frac{1}{1 + \left(\frac{2-a}{a}\right)\left(\frac{\gamma}{1+\gamma}\right) \frac{4}{N_{Pr(s)}} K_n} \right]^{0.45} \tag{23}$$

Where, a the thermal accommodation coefficient
γ the specific heat ratio (c_p/c_v)
$N_{Pr(s)}$ the Prandtl number of the gas at the surface temperature of the particle

Typical numerical values of the correction factor in square brackets can vary between 1.0 to 0.4, with an increase of the Knudsen number from 0.01 to 1.0. An example is given in Fig. 16 from [23] for an Ar-H_2 plasma (mole ratio 1:4), an accommodation coefficient a = 0.8 and particles surface temperature of 1000 K.

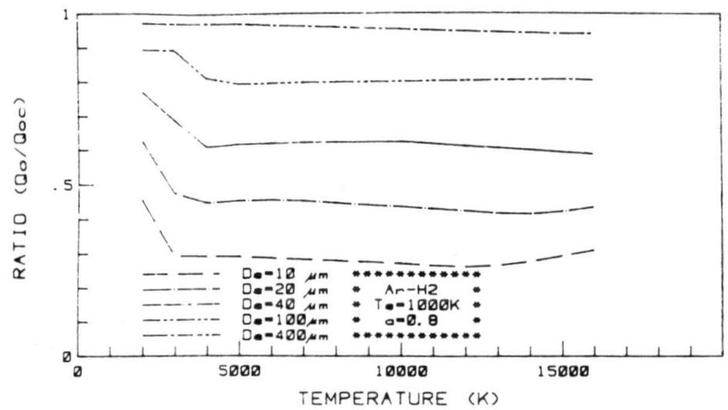

Figure 16: Ratio of the heat flux with and without Knudsen effect for particles of different size in an Ar-H_2 plasma (mole ratio 1:4, T_s = 1000 K, a = 0.8).

The ratio Q_o/Q_{oc} decreases when the accommodation coefficient a decreases and one of the main problem is the proper choice of the value of a . It is also worth to notice that the Knudsen effect is stronger for plasmas with higher enthalpies [23].

A corresponding correction to account for non-continuum effects in plasma particle heat transfer over the same Knudsen number range ($0.01 < Kn < 1.0$) has also been proposed by Chen and Pfender [24]. In this case they made use of the temperature jump concept resulting in the following correction;

$$q = (q)_{cont} \left[\frac{1}{1 + (2Z^*/d_p)} \right] \tag{24}$$

Where, $(q)_{cont}$ is the heat flux calculated using available correlation for continuum flow and Z^* is the jump distance.

Typical numerical values of the correction factor in square brackets in this case could be as low as 0.4 or 0.5 for a Knudsen number of about 0.1.

3.3.3 Influence of the heat conduction in the particle

As will be shown later, one of the important problems facing most research workers dealing with the calculations of trajectories and temperature histories of particles injected into a plasma is how to decide, a priori, whether internal conduction in the particle should be taken into account or not.

Intuitively, it should always be possible to treat small particles of diameters in the range of 10 to 20 μm as iso-thermal and neglect the effect of internal heat conduction on the overall heat transfer process. This, however, has been shown (see eq. 13) to depend on the Biot number. When heat propagation has to be taken into account, the time evolution of the temperature field within the particle is given by the following conduction equation:

$$\frac{1}{r^2}\frac{\partial}{\partial r}\left(\kappa_p r^2 \frac{\partial T}{\partial r}\right) = \rho_p c_{pp} \frac{\partial T}{\partial T} \qquad (25)$$

where the index p refers to the particle and r is the radial distance from the center of the particle. The velocity of propagation of a melting front $\left(\frac{\partial r}{\partial t}\right)_{rm}$ is given by;

$$\left(\frac{\partial r}{\partial t}\right)_{rm} = \frac{I}{\rho_l \Delta H_m}\left[\kappa_{pl}\left(\frac{\partial T}{\partial r}\right)_{rm} - \kappa_{ps}\left(\frac{\partial T}{\partial r}\right)_{r_m}\right] \qquad (26)$$

where κ_{pl} and κ_{ps} are the thermal conductivities respectively of the liquid and solid phases, r_m is the radius of the melting front and ΔH_m the latent heat of melting. The boundary conditions for equ. (25) and (26) are summarized in Fig. 17.

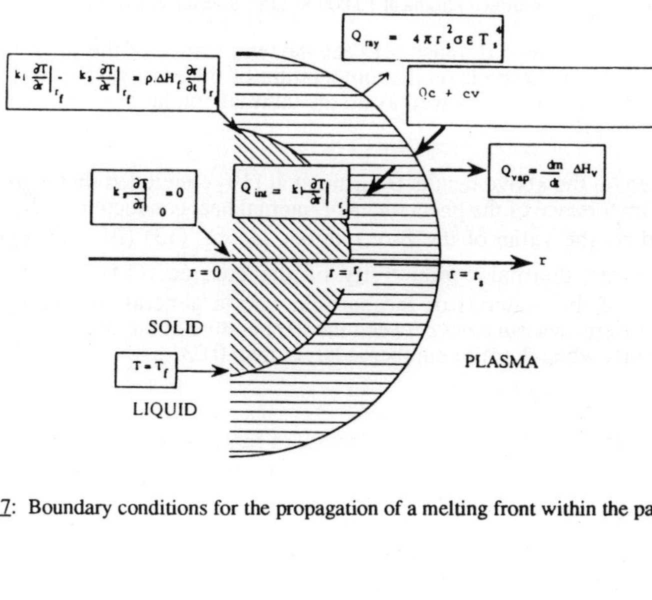

Figure 17: Boundary conditions for the propagation of a melting front within the particle.

Bourdin et al [19] carried out a systematic study of the transient heating of spherical particles of different materials (Ni, Si, Al_2O_3, W, SiO_2), of different particle sizes (d_p = 20-400 μm) as they are suddenly immersed in different plasmas (Ar, N_2, H_2) at atmospheric pressure and different temperatures (T_∞ = 4 000 - 10 000 K).

Typical results obtained for alumina particles of different diameters in a nitrogen plasma at 10 000 K are given in Fig. 18. This figure shows temperature differences between the surface (dotted line) and that of the center of the particle (solid line) as high as 1 000 K even for the 20 μm diameter particles.

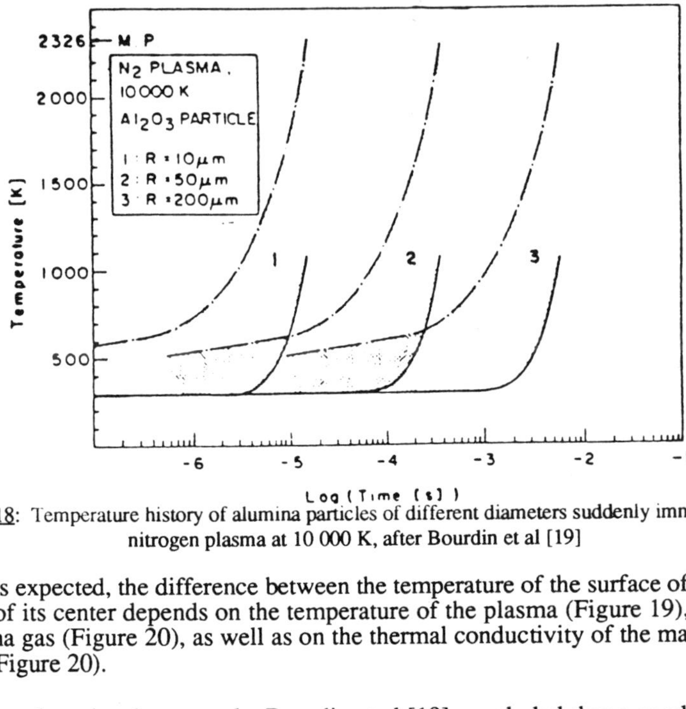

Figure 18: Temperature history of alumina particles of different diameters suddenly immersed in a nitrogen plasma at 10 000 K, after Bourdin et al [19]

As expected, the difference between the temperature of the surface of the particle and that of its center depends on the temperature of the plasma (Figure 19), the type of the plasma gas (Figure 20), as well as on the thermal conductivity of the material of the particle (Figure 20).

Based on the above results Bourdin et al [19] concluded that a good estimate of the relative importance of the phenomena of internal heat conduction in the particle can be made based on the value of the Biot number see eq. (13) ($Bi = \overline{\kappa} / \kappa_p$) with, $\overline{\kappa}$, the integrated mean thermal conductivity of the gas, eq. (14), and κ_p the thermal conductivity of the material of the particle. As a general rule they observed that substantial differences can exist between the temperature at the surface of the particle and that at its center when the Biot number is larger than 0.03.

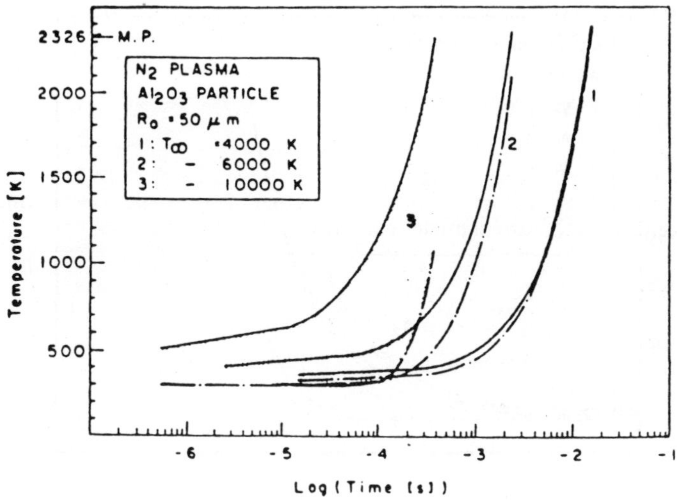

Figure 19: Temperature history of 100 μm diameter alumina particles immersed in nitrogen plasma at different temperatures, after Bourdin et al [19].

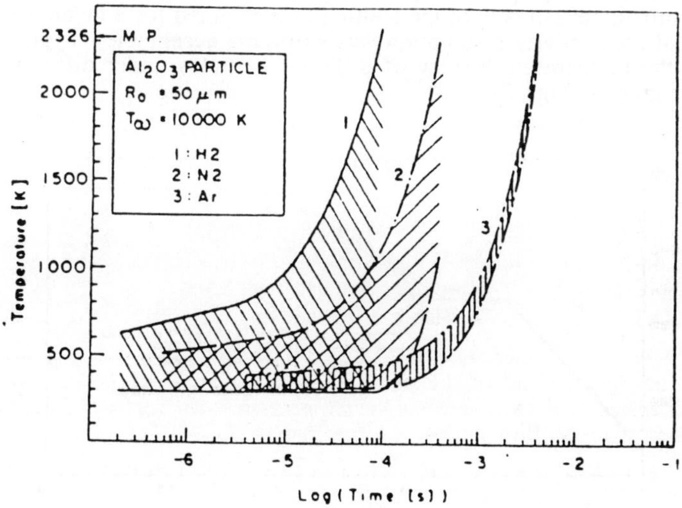

Figure 20: Temperature history of 100 μm diameter alumina particles immersed in different plasmas at 10 000 K, after Bourdin et al [19]

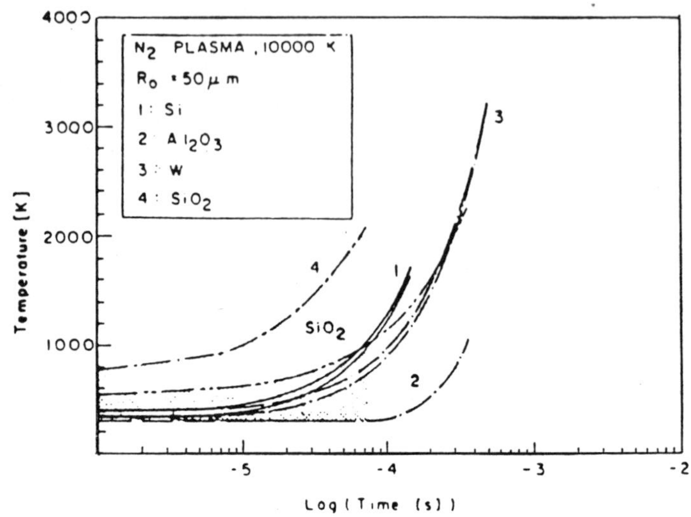

Figure 21: Temperature history of 100 μm diameter particles of different materials immersed in a nitrogen plasma at 10 000 K, after Bourdin et al [19].

A similar conclusion was reached by Chen and Pfender [25] in their study of the unsteady heating of small particles in a thermal plasma. Their results showed, however, that for an approximate estimate of the melting time required for a given particle, the infinite thermal conductivity assumption was still quite acceptable. Typical results, presented as the temperature history of different particles under different plasma conditions, are given in Figs. 22 to 24.

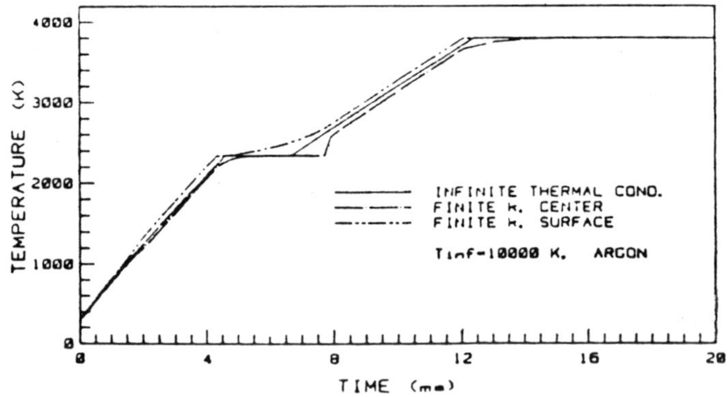

Figure 22: Temperature history of an alumina particle exposed to an argon plasma at 10 000 K (after Chen and Pfender [25]

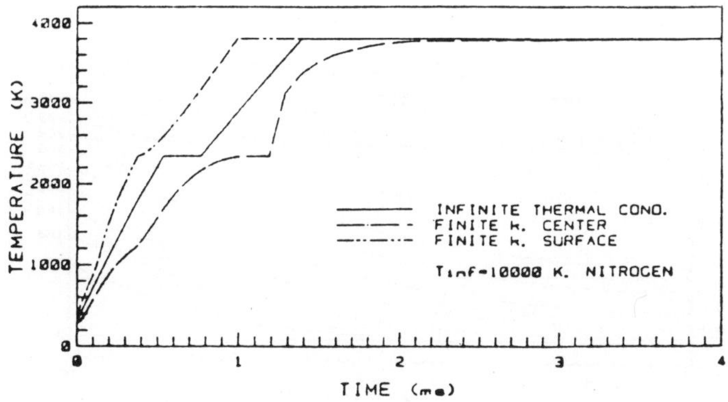

Figure 23: Temperature history of an alumina particle exposed to a nitrogen plasma at 10 000 K (after Chen and Pfender [25])

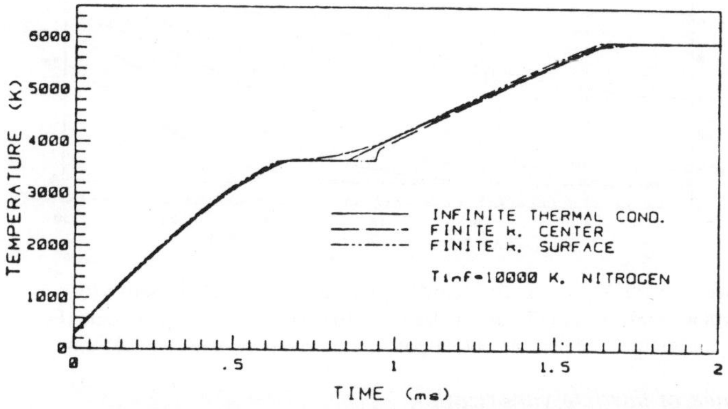

Figure 24: Temperature history for a tungsten particle exposed to a nitrogen plasma at 10 000 K (after Chen and Pfender [25]

The heat propagation phenomenon is particularly important for particles with a low effective thermal conductivity, such as agglomerated particles, which can have a porosity as high as 45% by volume, compared to dense fused and crushed (FC) powders. Typical results giving the evolution of the temperature profile in 60-µm diam. Zirconia (ZrO_2) particles immersed in an Ar/H_2 plasma at 10 000 K are given in Fig. 25. The profiles given in the top part of the figure were calculated for a fully dense particle (FC) while that at the bottom of the figure correspond to a porous particle (45% by vol.) with a mean pore diameter of 50 nm.

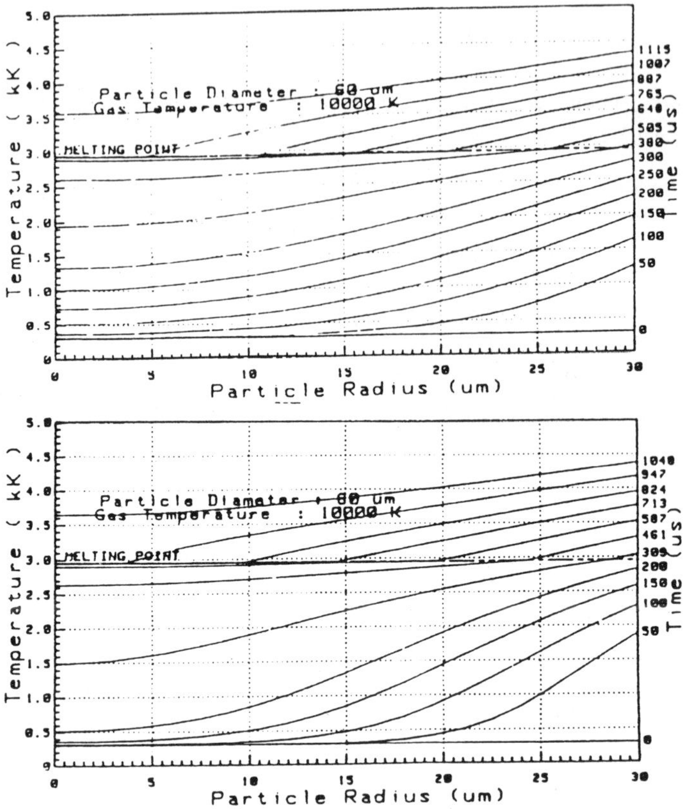

Figure 25: Effect of particle porosity on internal temperature gradients. Example for a 60 μm diameter ZrO_2 particle in an Ar/H_2 (17 vol %) plasma at 10 000 K top. - dense particle (FC) bottom agglomerated particle : (45% porosity), mean pore diameter 0.05 μm.

3.3.4 Influence of particle vaporization

Vaporization will start to take place as the particle surface temperature approaches its boiling point. The generated vapor cloud will surround the particle affecting the heat transfer rates between the plasma and the particle. This phenomenon will be particularly important with low thermal conductivity materials and high thermal conductivity plasmas such as for example ZrO_2 particles in Ar/H_2 plasma, or when treating low boiling temperature materials even in low thermal conductivity plasmas.

a) Calculation of the evaporation rate

At the liquid-vapor interface there is a vapor pressure p which is such that the chemical potentials are the same within the liquid and the vapor. If the vapor pressure far away from the interface is lower than that of the saturated vapor p_v mass transfer from the surface liquid droplet to the vapor clouds will take place. The evaporation rate in a pseudo steady state is given [20, 27, 28] by:

$$\frac{dm}{dt} = S \cdot h_m \cdot \rho_g \cdot M \cdot \ln\left(\frac{p}{p - p_v}\right) \tag{27}$$

Where S is the surface area of the particle, h_m the mass transfer coefficient, ρ_g the vapor density in the boundary layer around the particle, M molecular weight of the evaporating material, p the saturation vapor pressure and p_v the vapor pressure at the surface temperature of the particle. The mass transfer coefficient is calculated using the Ranz - Marshall [29, 30] correlation, if necessary using a correction factor of the same type of that used for the Nusselt number (see eq. (18 to 20)), through the Sherwood number ($Sh = h_m \, d_p / D$)

$$Sh = 2 + 0.6.\ Re^{1/2}\ Sc^{1/3} \tag{28}$$

where D is the diffusion coefficient and $Sc = \mu / \rho D$. Under conditions of high mass transfer rates, the particle vaporization process can be heat transfer controlled. The particle vaporization rate in this case is given by the following energy balance equation.

$$\frac{dm_p}{dt} = \frac{Q}{\Delta H_v} \tag{29}$$

where m_p is the mass of the particle and Q is the net heat exchanged between the plasma and the particle.

b) Heating of the vapor

At the interface between the liquid and vapor phases, a heat balance is maintained and at the same time a bulk flow of material crosses the interface. For high mass transfer rates, the transfer coefficients become functions of the mass transfer rate, thus causing non-linearity in the transport equations. For example, Chen and Pfender [31] have shown that the heat flux through a liquid-vapor interface is reduced due to absorption of heat by the vapor. The heat received by the particle can be written [31] as:

$$Q = 2\pi d_p \ \Delta H_v \int_{T_s}^{T_\infty} \frac{\kappa_g \cdot dT}{h - h_s + c} \tag{30}$$

with

$$c = \Delta H_v + \frac{2\pi \, d_p^2 \in \sigma_s \, T_s^4}{\frac{dm}{dt}} \tag{31}$$

where h is the plasma specific enthalpy at T, h_s being that at T_s. This expression has been deduced assuming that the plasma thermodynamic and transport properties (κ, h) are not modified by the vapor diffusion.

Assuming that $\kappa/c_p = \left(\overline{\kappa} \ / \ \overline{c_p}\right)$ in the boundary layer, eq. (30) applies because

$$Q = \frac{\Delta H_v}{h_\infty - h_s} \ln\left(1 + \frac{h\infty - hs}{\Delta H_v}\right) 2\pi d_p \left(\frac{\overline{K}}{c_p}\right)\left(h_\infty - h_s\right) \tag{32}$$

Eq. (32) is equivalent to the introduction of a correction factor, λ_c to the thermal flux due to the plasma (see eq. (12), i.e.

$$\lambda_c = \frac{\Delta H_v}{h_\infty - h_s} \ln\left[1 + \left(\frac{h_\infty - h_s}{\Delta H_v}\right)\right] \tag{33}$$

The correction factor proposed by Engelke [32] and Borgianni [33].

Typical results are given in Fig. 26 and 27 as the ratio of the heat flux to the particles in the presence of vaporization, Q_1, to that in the absence of vaporization, Q_o, as a function of the plasma temperature. These show a substantial reduction of the heat flux to the particle with the increase of the vaporization rate. The effect is more pronounced for an Ar/H_2 or a N_2 plasma compared to that for a pure Ar plasma.

c: Radiation of the vapor

During the evaporation of metallic particles radiative energy losses from the metal vapor cloud surrounding the particle cannot be neglected as that of the plasma. This is illustrated in Fig. 28 after Essoltani [34] showing that in the temperature range from 3 000 to 8 000 K the radiation losses of the vapor can be several orders of magnitude higher that those of a pure argon plasma.

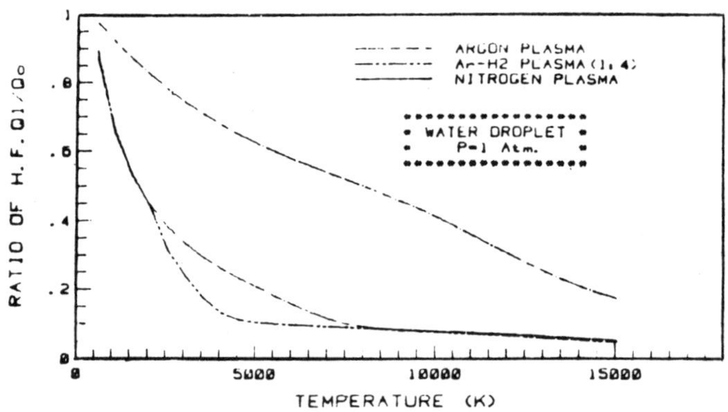

Figure 26: Effect to evaporation on heat transfer to a water droplet under different plasma conditions, after Chen and Pfender [31]

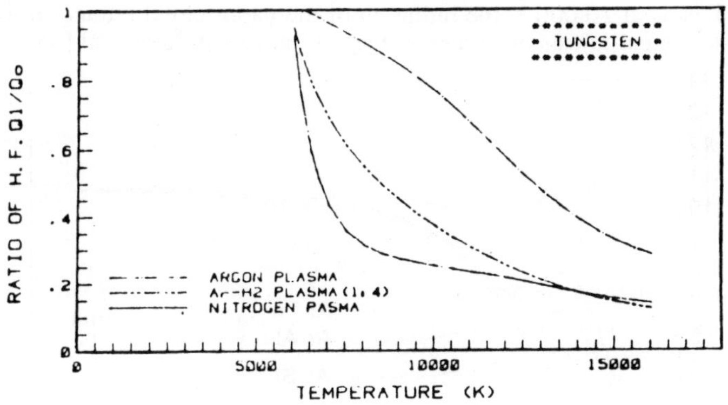

Figure 27: Effect of evaporation on heat transfer to a tungsten particle under different plasma conditions, after Chen and Pfender [31]

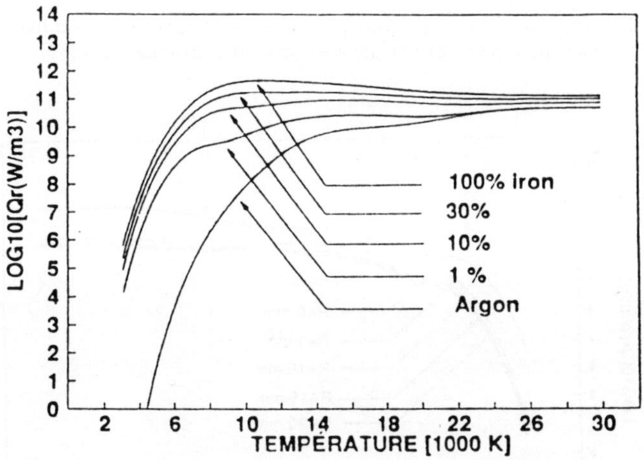

Figure 28: Volumetric radiation losses for an Ar/Fe plasma as function of temperature and the molar fraction of iron vapor, plasma radius R = 1 mm, after Essoltani et al [34].

The effect is observed with different metals, though its extent varies with the nature of the vapor as illustrated in Fig. 29 from Essoltani et al [35]. Of course the self absorption reduces the net radiation losses as illustrated in Fig. 30 for an Ar/Fe plasma, after Essoltani et al [34]. When calculating the heat transfer to an iron particle immersed in an argon plasma at 10 000 K the radiation of the vapor cloud reduces drastically the heat transfer to the particle as illustrated in Fig. 31 , after Essoltani et al [36].

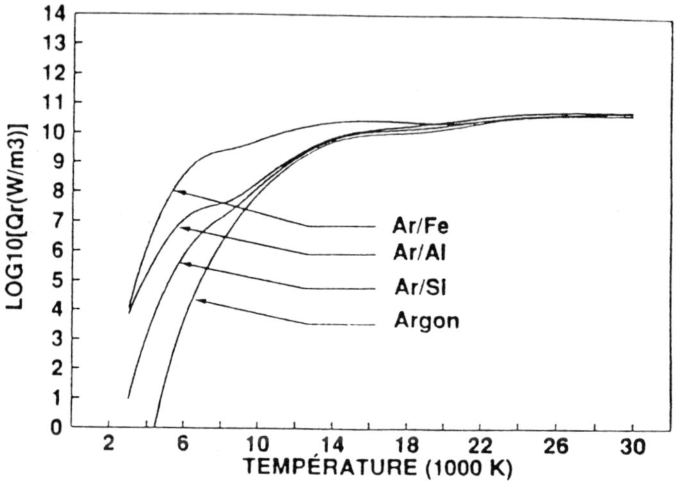

Figure 29: Volumetric emission for an argon plasma in the presence of 1% molar fraction of different metallic vapors, plasma radius = 1 mm, after Essoltani et al [35]

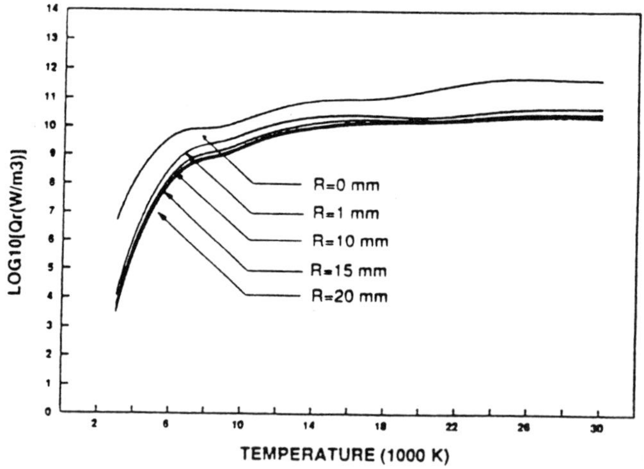

Figure 30: Self absorption effects for an Ar - Fe plasma as function of the temperature and the radius of the plasma [35].

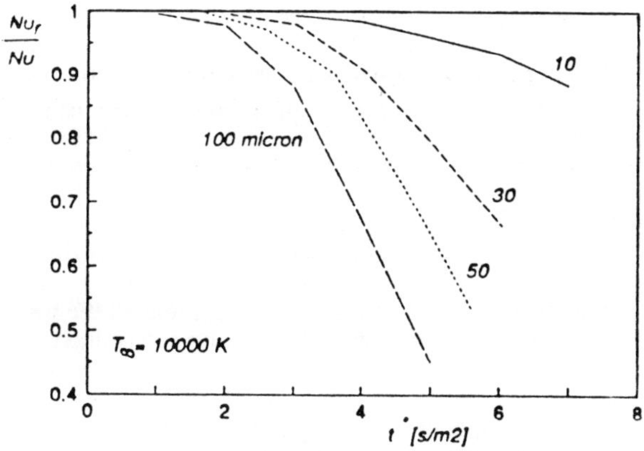

Figure 31: Reduced Nusselt number as function of time for different sizes of iron particles in an argon plasma at T_∞ = 10 000 K, after Essoltani et al [36]

3.3.5 Other effects

In all preceeding calculations, the particles have been assumed to be spherical. If this is the case for agglomerates, spray dried, atomized powders or those produced by sol-gel techniques, it is not at all true for fused and crushed particles which can have angular shapes with a low shape factor. The shape of the particle and its drag coefficient in this case can be correlated by the particle sphericity factor in a limited range of shape effects.

Particle charging can also have a limited effect on its trajectory since a particle injected into a thermal plasma will always assume a negative charge due to the difference between the thermal velocities and mobilities of electrons and ions. Whether this affects particles drag in thermal plasmas has never been explored. Under LTE conditions [10] in the boundary layer, particle charging will be of minor importance, because of the low charge concentration that can exist in the region near the particle surface. This may, however, be different for frozen chemistry, under non-continuum conditions in the case of low pressure plasma spraying.

Deviations from LTE conditions in the plasma can also be important, in particular close to the particles surface. Whether or not such deviations from LTE will substantially affect heat and momentum transfer to particles in the condensed phase remains still a matter for further study.

3.4 Momentum, heat and mass transfer under plasma conditions

As already mentioned in the previous sections the diffusion of the particle vapor in the plasma surrounding it has been neglected as well as its effect on the corresponding plasma transport properties. When such phenomena are to be taken into account, the calculation becomes far more complex [37].

3.4.1 No chemical reactions between vapor and plasma

In this case, the following phenomena governs the transport processes:

- Mass conservation which according to the perfect gas law can be expressed as function of the pressure.
- The conservation of the chemical species taking into account the diffusion transport of the each specie.
- The momentum conservation.
- The energy conservation including the transport of energy due to diffusion.

What makes the problem very complex is the fact that all the transport properties (including the different diffusion coefficients) have to be determined at each step of the calculation.

At a given time the composition of the gas at the interface (supposed to be at the particle surface temperature) has first to be defined in order to be able to calculate the thermodynamic and transport properties of the gas. The mass conservation equation allows to calculate the molar barycentric velocity, the momentum equation allows to determine the radial distribution of the pressure p(r), the conservation of the chemical species equation allows to determine the molar fraction of the different species and finally the energy equation gives the radial temperature distribution T(r). The latter can then be used to recalculate a new value of the composition at the interface up to the convergence. For example Fig. 32, after [37] represent the molar fractions of Si vapor, H_2 and H in the vapor clouds surrounding a 100 μm diam. silicon particle immersed in an H_2 plasma at 5 800 K.

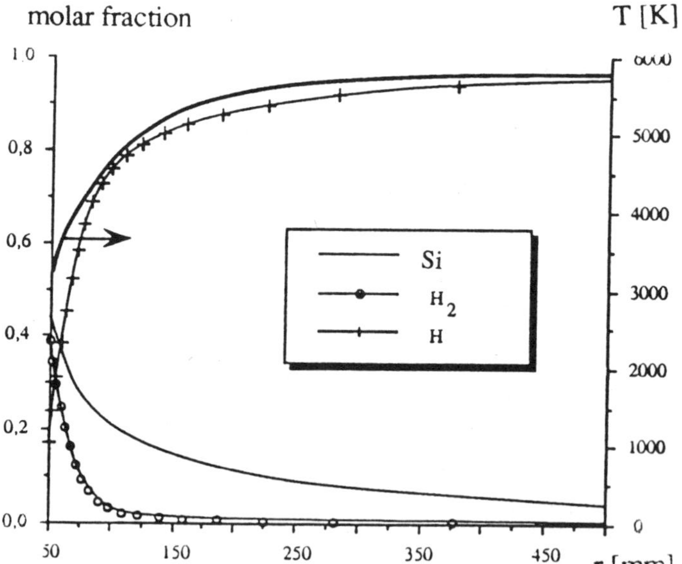

Figure 32: Temperature profile and molar fraction of Si, H_2 and H in the vapor clouds surrounding a 100 μm diam. Si particle immersed in an H_2 plasma at 5 800 K, after Hubert [37].

3.4.2 Chemical reactions between vapor and plasma

In this case the calculation is even more complex with the introduction of a source term representing the variation of the mole numbers of the different chemical species in the conservation equations. Typical results are illustrated in Fig. 33 from [37] for a SiO_2 particle, 50 μm in diameter, immersed in an argon/hydrogen plasma (20 vol %) H_2 at 5 800 K.

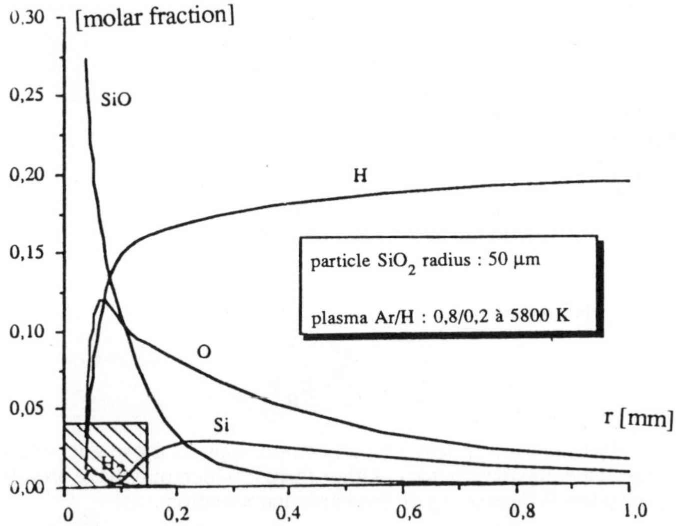

Figure 33: Evolution of the molar fraction of the different chemical species around a 100 μm diam. SiO_2 particle in an infinite Ar/H_2 (20 vol %) plasma at 5 800 K, after Hubert [37].

4. Single particle trajectories and temperature history

4.1 Introduction

In this section we will deal mainly with the calculation of single particle trajectories and temperature histories under plasma conditions. Included will be a discussion of the basic assumptions, the governing equations and typical results obtained for d.c. plasma jets and r.f. induction plasma discharges. For simplicity, these will be limited to low particle loading. The effect of plasma-particle interactions under dense loading conditions will be dealt with separately in the next section.

4.2 Basic assumptions and governing equations

During the displacement of a particle in a plasma stream, it is subjected to a number of forces which act simultaneously on it and have varying influence on its trajectory and residence time in the plasma. Among the most important and invariably taken into account are the inertia forces, F_i, and the viscous drag forces, F_D, which are defined as:

$$F_i = \frac{\pi d_p^3}{6} \rho_p \frac{dU_p}{dt} \tag{34}$$

$$F_D = \frac{\pi d_p^2}{4} C_D \frac{1}{2} \rho U_R^2 \tag{35}$$

Where,

d_p particle diameter
ρ_p particle density
ρ plasma density
C_D drag coefficient
U_p particle velocity
U_R relative velocity between the plasma and the particle
t time

The gravity force, F_g defined as

$$F_g = \frac{\pi d_p^3}{6} \rho_p g \tag{36}$$

can also be relatively important in low-velocity flows such as those encountered in r.f. discharge and large scale plasma reactors. Other forces, which may have an influence on the trajectory of the particle depending on the particular situations, are;

F_p The pressure gradient force which can be important in the presence of steep pressure gradients in the flow field.

F_B The Basset history term which can be important under very high acceleration conditions.

F_T Thermophoresis effects which can be important for very fine particles $d_p < 0.1$ μm in the presence of steep temperature gradients.

F_C Coriolis forces, which result from the rotation of the particles about an axis parallel to the direction of relative motion.

Considering that in most cases encountered in the melting and processing of powders under plasma conditions, F_p, F_B, F_T, and F_C are negligible, a force balance around a single particle in motion in a plasma flow can be written simply as;

$$F_i = F_D + F_g \tag{37}$$

Substituting Eqs. (34), (35) and (36) in Eq. (37), the latter can be written in axis symmetric cylindrical coordinates as follows;

$$\frac{d\,U_p}{dt} = -\frac{3}{4}\,C_D\,(U_p - U)\,U_R\left(\frac{\rho}{\rho_p\,d_p}\right) \pm g \tag{38}$$

$$\frac{dV_p}{dt} = -\frac{3}{4}\,C_D(V_p - V)\,U_R\left(\frac{\rho}{\rho_p\,d_p}\right) \tag{39}$$

The + or - sign in front of the gravitational acceleration, g, corresponds to a flow in axial direction with the positive direction for the velocity of the particles being downwards or upwards, respectively.

As for the instantaneous particle temperature, this can be calculated by an energy balance on the particle (using eq. (12)). Assuming a particle of infinite thermal conductivity (i.e. uniform particle temperature $T_p = T_s = T_c$), the variation of the particle temperature during its trajectory can be calculated by any of the following equations depending whether the particle is undergoing a phase change or not.

For $T < T_m$ or $T_m < T < T_v$:

$$\frac{dT_p}{dt} = \frac{6\,Q_n}{\pi d_p^3\,\rho_p\,c_{ps}} \tag{40}$$

Where T_m and T_v are the melting and boiling temperatures of the particle material, respectively, and c_{ps} is the specific heat of the particle.

For $T=T_m$, the particle undergoes a phase change with the mass fraction which is molten, x, given by Eq. (41) as follows,

$$\frac{dx}{dt} = \frac{6\,Q_n}{\pi d_p^3\,\rho_p H_m} \tag{41}$$

Where, H_m is the latent heat of fusion of the particle material.

For, $T \approx T_v$, the particle evaporates with the rate of change of the particle diameter given by Eq. 42 as :

$$\frac{d\,d_p}{dt} = \frac{-2\,Q_n}{\pi d_p^2\,\rho_p\,H_v} \tag{42}$$

Where, H_v is the latent heat of vaporization of the particle material.

It should be pointed out that Eq. 42 is only an approximation and that evaporation is likely to take place well below the boiling point of the material. As shown by Chen and Pfender [31] the shielding effect of the particle by the evolving vapor cloud could lead to considerable slowing down of the heating rate of the liquid droplet with the possibility of complete evaporation of the particle before it ever reaches its boiling temperature (see section 3.3.4). Interestingly, they also show that the phenomena does not seem to affect to any considerable extent the estimate of the time required for the complete evaporation of the particle.

4.3 Typical results of case studies

Typical results obtained by different investigators [38 to 48] who worked on the mathematical modelling of the processing of different powders under plasma conditions will be given as examples of some of the present modelling capabilities.

4.3.1 Particle trajectory calculations in an air jet by Lesinski et al [44]

In this case computations were carried out for a relatively simple flow configuration which is similar to that used in the d.c. plasma spraying of powders except that the plasma jet was replaced by a turbulent air jet at ambient temperature and the results compared with experimental data obtained using laser Doppler anemometry (LDA) under similar operating conditions.

The computed particle trajectories for glass beads of different particle diameters, injected at different initial velocities, at right angle to the air jet are given in Fig. 34 The center line velocity of the jet at the exit of the nozzle was 95 m/s. Radial velocity profiles obtained for glass beads with a mean particle diameters of 97.0 μm and a standard deviations of 10.0 μm injected at a mean velocity of 5.0 m/s in the same air jet are given in Fig. 35. Superimposed, for comparison, are LDA experimental data which agree remarkably well with the computed mean particle velocity profiles. The experimental data, however, show a wider spread of the particles in the jet which could be attributed to the stochastic nature of the process involved, to the wide dispersion (alone angle up to 30°) of the particles at the injector exit (see section 2) which is very difficult at the moment to introduce in the descretization of the particle injection velocities and thus in the particle trajectories.

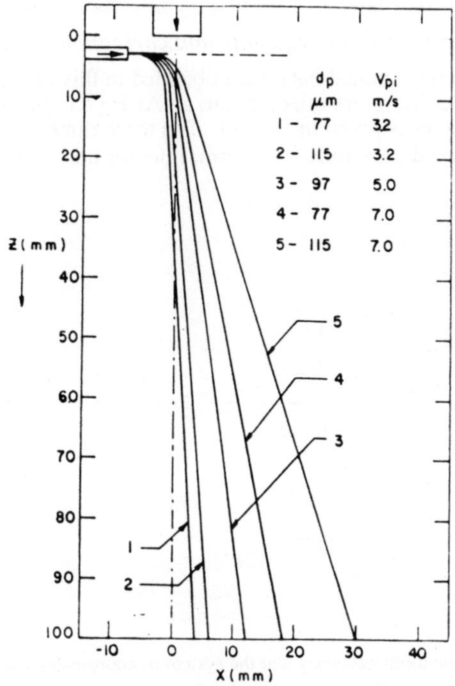

Figure 34: Calculated trajectories for single glass beads injected in a turbulent air jet (u_0 = 95 m/s), after Lesinski et al [44]

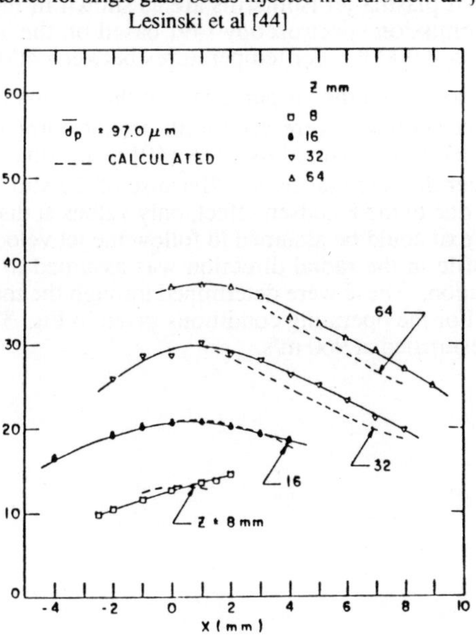

Figure 35: Mean particle velocity for 97.0 μm mean diameter glass beads injected with an average velocity of 5.0 m/s in a turbulent air jet, after Lesinski et al [44]

4.3.2 Particle trajectory and temperature history in a d.c. plasma jet

a) Influence of temperature gradients and correction factors

Theoretical and experimental data were obtained in this case for alumina particles of different mean particle diameters injected into an Ar/H_2 d.c. plasma jet. A schematic sketch of the torch geometry is given in Fig. 36. The torch dimensions and the operating conditions are as follows: $d_o = 8$ mm, $\delta = 2$ mm, injector i.d. 1.8 mm

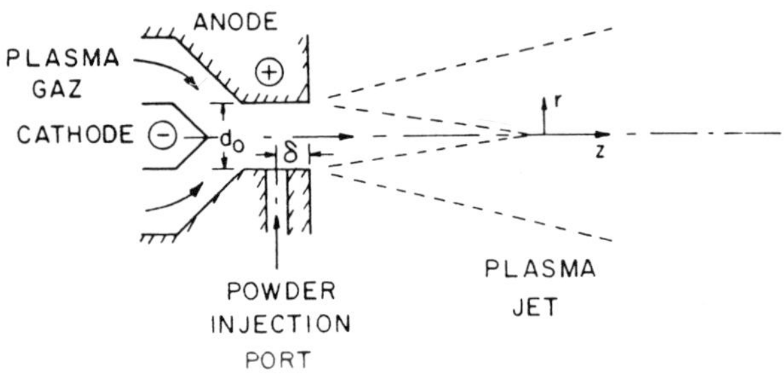

Figure 36: Schematic of the torch geometry and the system of coordinates after Vardelle et al [46].

The measured plasma jet isotherms are as shown in Fig. 37. The temperatures were measured by emission spectroscopy [49] based on the absolute intensity of Ar I 727.2 nm line for T > 8 000 K. For temperatures between 4000 and 6000 K, rotational bands of N_2^+ were used (from the air pumped into the plasma jet) while at temperatures below 3000 K, thermocouples were used with proper correction for radiation losses. The velocity of the jet was measured by LDA [49] using small alumina particles $d_p < 1$ µm) injected as tracer into the plasma jet. Because of the slip between the particles and the plasma velocity due to the Knudsen effect, only values at distances higher than 25 mm from the jet exit could be assumed to follow the jet velocity. For lower distances, the jet velocity profile in the radial direction was assumed to be similar to that of the temperature distribution. These were determined through the mass flowrate and enthalpy conservation [49]. For the operating conditions given in Fig. 37, the maximum velocity a the nozzle exit is estimated at 960 m/s.

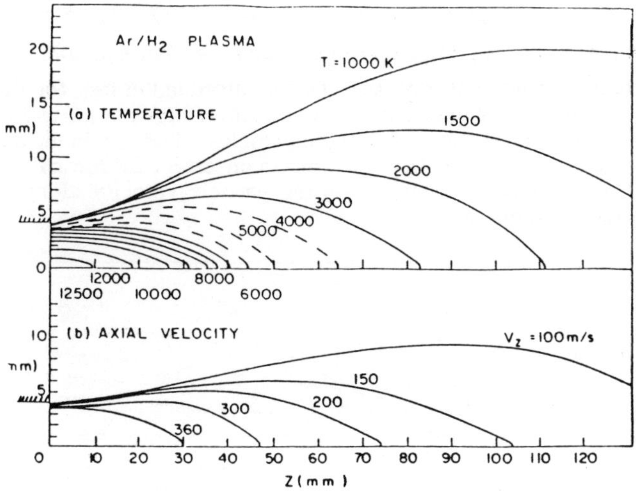

Figure 37: Temperature and axial velocity isocontours for an Ar/H_2 d.c. plasma jet. Plasma gas 75 slm Ar + 15 slm H_2, power = 29 kW, η = 63%, after Vardelle et al [46].

It has to be kept in mind, however, that the trajectory of a single particle, of a given mass and injection velocity, depends on the choice of the correction to the drag coefficient for the steep temperature gradients (see section 3.3.4), for the non-continuum effect, (see section 3.3.2) and for the particle vaporization (see section 3.3.4). This is illustrated in the following figures for 60 µm diam. alumina particles (ρ_p = 3 900 kg/m^3) injected at right angle to the axis of the plasma jet depicted in Fig. 37. The particle trajectories given in Fig. 38 were obtained using three different expressions for drag coefficient correction. It can be readily seen that large differences occur which are particularly important when dealing with smaller particles.

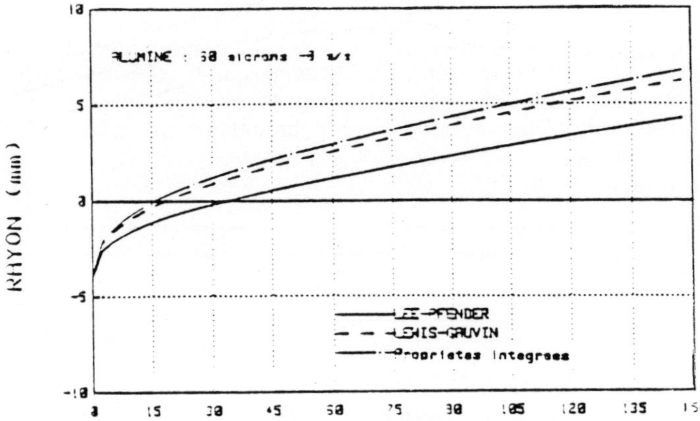

Figure 38: Effect of the drag correction coefficient on the trajectories of 60 µm diam. alumina particles injected at 8 m/s into the plasma jet depicted in fig. 37 : Lee et al [14], Lewis and Gauvin [13], integrated properties [19].

Such differences in the particle trajectories will obviously be reflected on the corresponding difference of the particle temperature histories. Once the correction for the steep gradients has been chosen, the corrections for non-continuum effects and vaporization have also to be determined. This is illustrated in fig. 39a for 20 μm diam. alumina particles (to have a significant non-continuum effect) injected at 25 m/s (to obtain about the same trajectory as that of the 60 μm particles). Here again the different type of corrections used are shown to have an important effect on the calculated particle trajectory. The corresponding effect on the particle velocity along each of these trajectories is shown in Fig. 39b.

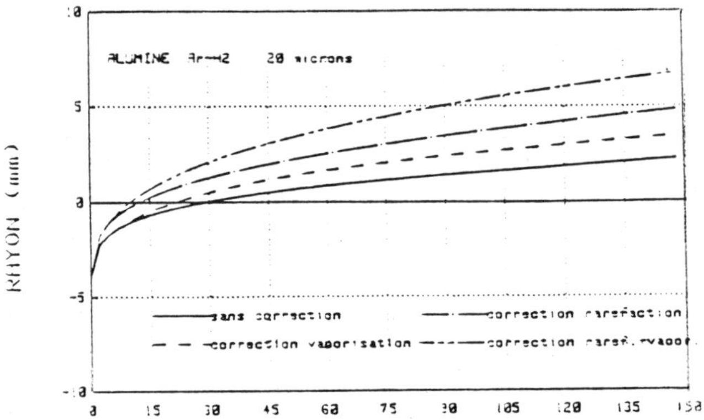

Figure 39a: Trajectories of 20 μm in diam. alumina particles injected at 25 m/s in the plasma jet depicted in figure 37. The temperature gradients correction has been calculated using the mean integrated properties.

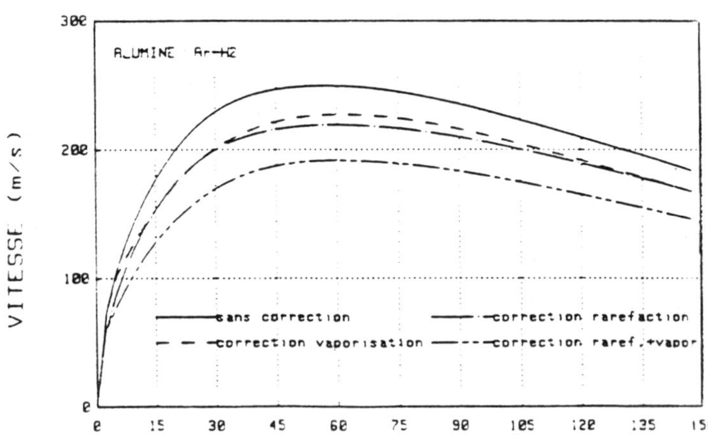

Figure 39b: Particle velocity along the trajectories represented in figure 39a correspondingly, different drag coefficient corrections for non-continuum effects and particles vaporization.

The effect of turbulence on the particle motion has also been shown to have an important influence on the particle trajectory by Pfender et al [15]. Typical results for 10 to 50 µm alumina particles injected into an argon d.c. turbulent plasma jet are given in Fig. 40.

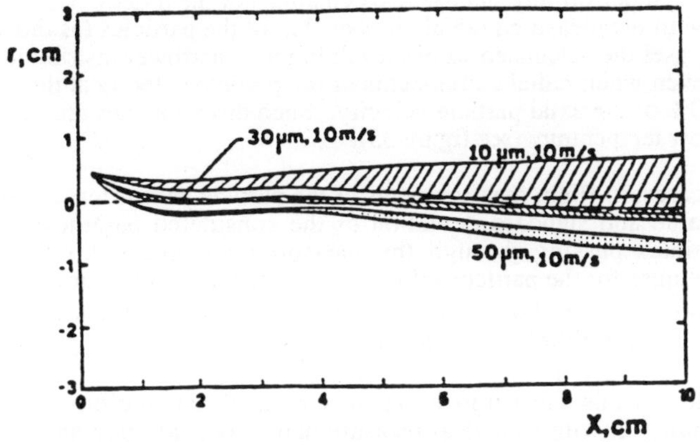

Figure 40: Dispersed alumina particle trajectories in an argon, d.c. turbulent plasma jet [15].

The ranges of dispersed particle trajectories are obviously larger for the smaller particles (10 µm). This is due to the small mass of such particles which follow easier the motion within the eddies. It is also observed that the dispersion becomes larger downstream. This is caused by the accumulated "random walk" influence. Comparing Fig. 40 to Fig. 38 and 39 it seems that the different corrections by the equations of motion may still exceed the influence of turbulence, for particles larger than 15 µm in diameter.

b) Heat propagation phenomenon

As soon as the Biot number (see equation (13) and section 3.3.3) is higher than 0.03 the heat propagation in the particle can no longer be neglected. This is illustrated in Fig. 41 for alumina particles, respectively 20, 40 and 60 µm injected in an Ar-H_2 d.c. plasma jet. Fig. 41 represents the velocities along their mean trajectories and the surface and center temperatures of the particles. These were injected in the plasma jet with different velocities in order to follow as closely as possible the same trajectory.

As could be expected (see figure 41a), the smallest particles reach the highest velocity but they decelerate faster than the bigger ones. In spite of a ratio of almost 2 in the residence times of the 60 µm particles compared to the 20 µm ones, their surface temperature is almost 1 200 K lower. The heat propagation phenomenon is also more important with the big particles. When considering 40 µm particles, the center temperature reaches that of the surface only 80 mm downstream of the nozzle exit.

c) Comparison between calculations and measurements

This comparison is always difficult for the following reasons. First, as underlined in section 1 the particle trajectories depend on there size distribution and injection velocity distribution (with non negligible radial components resulting in a dispersion cone angle up to 30° or even more). The injection velocity distribution is not very well known as well as the matching of this distribution with the corresponding particles size. When performing stochastic calculations of the trajectories, the results do not correspond with the measured radial number flux of the particles (as those shown in fig. 6). In most cases the calculated values result in much narrower distributions than the measured ones even when radial components of the particle velocity at the injector exit are as high as 30% of the axial particle velocity. Such discrepancies are observed even for an air jet at low temperature (see figure 35).

In most cases, in-flight measurements of the particle parameters are statistical in nature (mean value and standard deviation of the considered parameter for a given number of particles passing through the measurement volume). Moreover, the measurement volumes for the particle velocity v_p, diameter d_p and surface temperature T_S are not always the same. For example in general the measurement volume for, T_S, is much larger than that for the other parameters.

For these reasons the comparison of statistical calculations, such as those depicted in section 5.2, with statistical measurements does not give any better results than the comparison of the calculations for a single particle with the mean diameter of the size distribution traveling along the mean trajectory (defined as in section 1). As such a calculation is by far simpler and faster than any stochastic calculation, it is used in most cases.

The velocity of alumina particles of different sizes injected into the plasma jet depicted in fig. 37 with injection velocities such as to have about the same mean trajectory, was calculated and the mean velocity of particles with narrow size distributions are presented in Figures 42 and 43. The results show a reasonable agreement between the computed velocities for a single particle and those experimentally determined using laser anemometry. The results show generally a good agreement between the computed and measured particle velocities. The computed particle surface temperature, however, while being close to the experimental values in the early part of the trajectory, show a considerably faster particle cooling than that experimentally observed. The difference could be related to uncertainties in the prediction of radiation heat losses from the particles, which can be important in this temperature range.

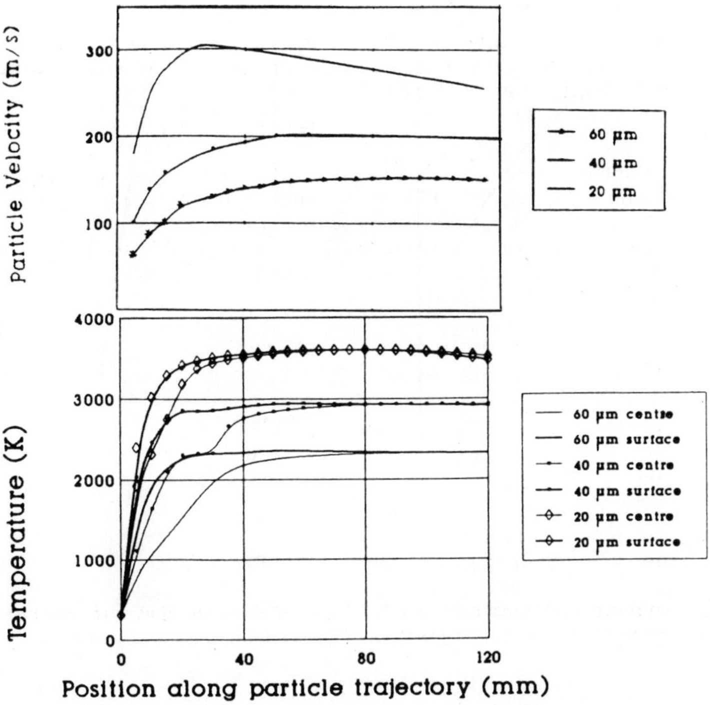

Figure 41: Particle velocity and temperature history for dense alumina particles injected in an Ar - H_2 d.c. plasma jet: nozzle i.d. 6 mm, 35 slm Ar, 12 slm H_2, I = 600 A, P = 35 kW, ρ = 55%.

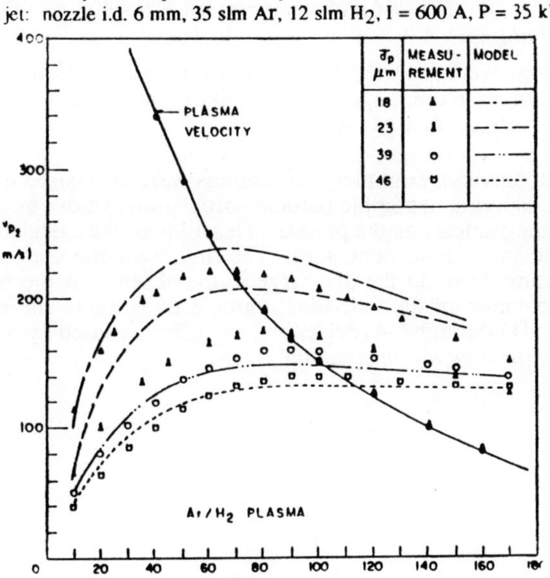

Fig.42: Particle velocity along their mean trajectory in the Ar-H_2 plasma jet depicted in Fig. 37 after Vardelle et al [46].

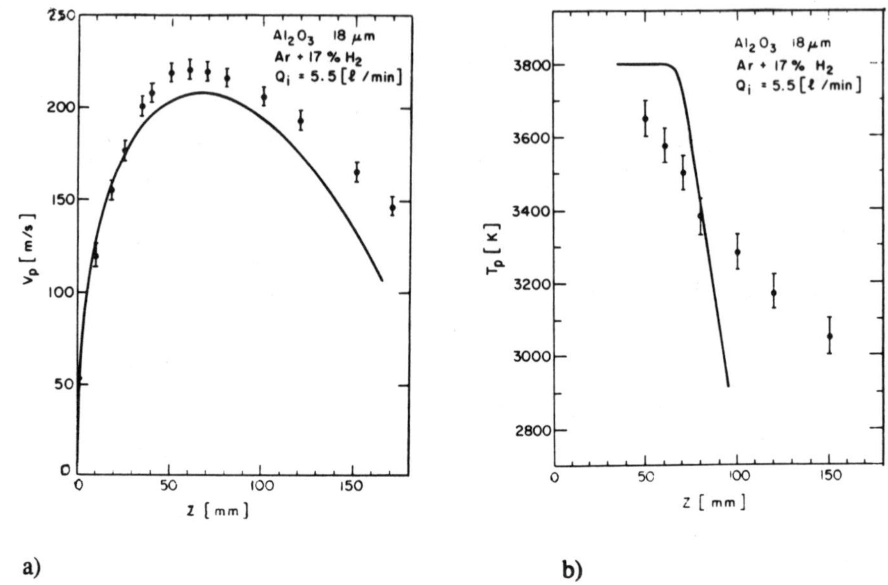

Figure 43: Experimental and computed velocity (a) and surface temperature (b) of alumina particles along their mean trajectory in the plasma jet depicted in Fig. 37, after Vardelle et al [47].

d) Particle trajectory and their temperature history in an r.f. inductively coupled plasma

Computations were reported by Boulos [51] for fine alumina particles with a particle size distribution in the range of 10 to 200 μm, injected axially in the discharge region of an inductively coupled r.f. plasma.

As with the previous examples, calculations were also based on the assumption of a dilute system implying negligible particle-particle interactions as well as negligible loading effects of the particles on the plasma. This allowed the calculations to be carried out in two separate and independent steps. The first involving the computation of the flow and temperature fields in the discharge in the absence of the particles. This is followed by computation of the individual particle trajectories and their temperature history using equ. (41) through (44) neglecting any effects caused by the presence of the particles on the plasma flow and temperature fields.

Typical results obtained for an argon plasma at atmospheric pressure are given in figures 44 and 45 for the flow and temperature fields, and the particle trajectories respectively. The plasma confinement tube had an internal diameter of 28 mm. The central powder carrier gas, intermediate and sheath gas flow rates were set to 0.4, 2.0 and 16.0 slm, respectively. The oscillator frequency was 3 MHz and the net power dissipated in the discharge was 3.77 kW.

In the representation of the trajectories given in figure 45, the size of the circles used to indicate the position of the particles is also an indication, although not to scale, of their diameter. Moreover, an open circle indicates a solid particle while a dark circle represents a liquid droplet. It is obvious that the predictions which could be made using such a model can be very useful when it comes to the optimization of the injection condition of a given powder with the objectives of obtaining either a physical and/or chemical change in the powder.

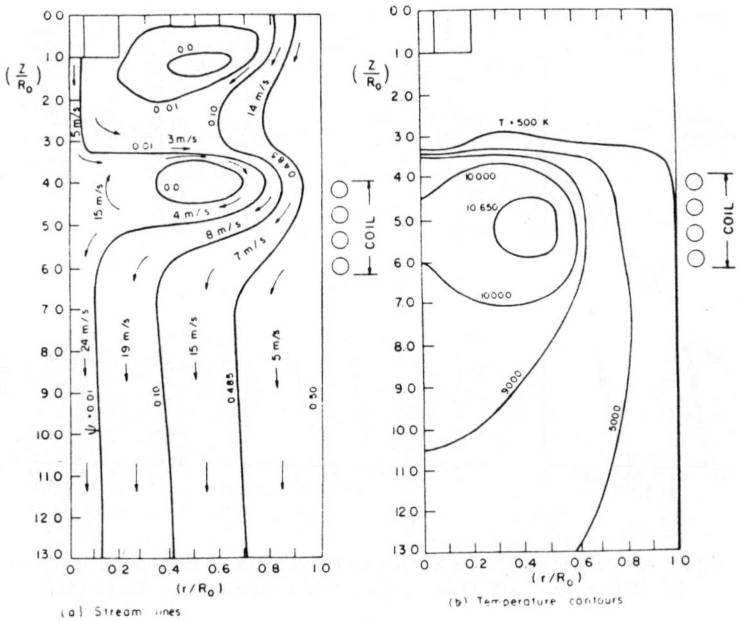

Figure 44: Flow and temperature fields in an Argon inductively coupled r.f. plasma, power = 3.77 kW, Q = 20 slm, Ro = 14 mm, after Boulos [51].

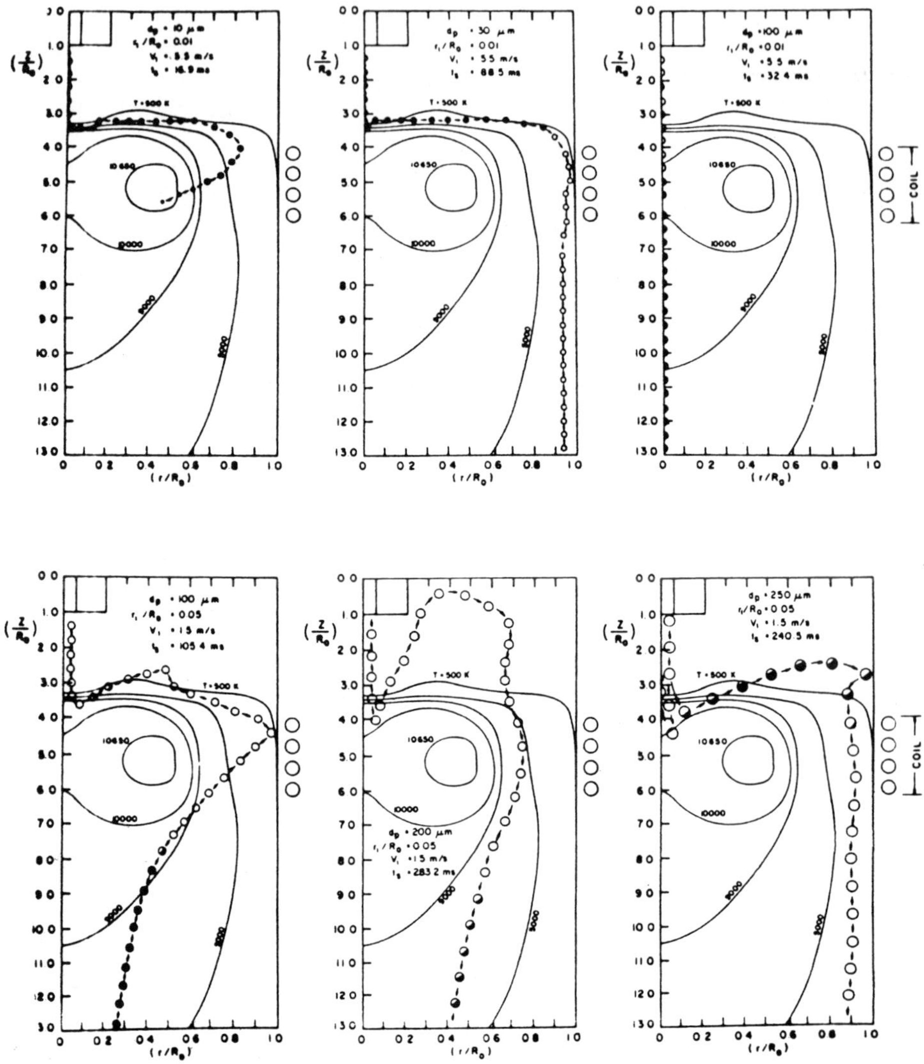

Figure: 45: Particle trajectories for alumina powder injected in an induction plasma f=3 MHz, P_t=3.77 kW, Q_1=0.4, Q_2=2.0, Q_3=16.0 slm, after Boulos [51].

5. Plasma-particle interactions under dense loading conditions

5.1 Introduction

While the assumption of a dilute system has generally been accepted for the calculation of individual particle trajectories and temperature histories under plasma conditions, the interpretation of the results obtained is greatly hindered by the simple fact that any application of plasma technology for the in-flight processing of powders will have to be carried out under sufficiently high loading conditions in order to make efficient use of the thermal energy available in the plasma. With the local cooling of the plasma due to the presence of the particles, model predictions using the low-loading assumption can be substantially in error.

In an attempt to take into account the plasma-particle interaction effects, Boulos and his collaborators [52-55] developed a mathematical model which through the interactive procedure illustrated in figure 46 up-dates continuously the computed plasma temperature, velocity and concentration fields. The interaction between the stochastic single particle trajectory calculations and those of the continuum flow, temperature and concentration fields is incorporated through the use of appropriate source-sink terms in the respective continuity, momentum, energy and mass transfer equations. These are estimated using the so-called particle-source-in-cell model (PSI-Cell) [56] which can be represented schematically as shown in figure 47.

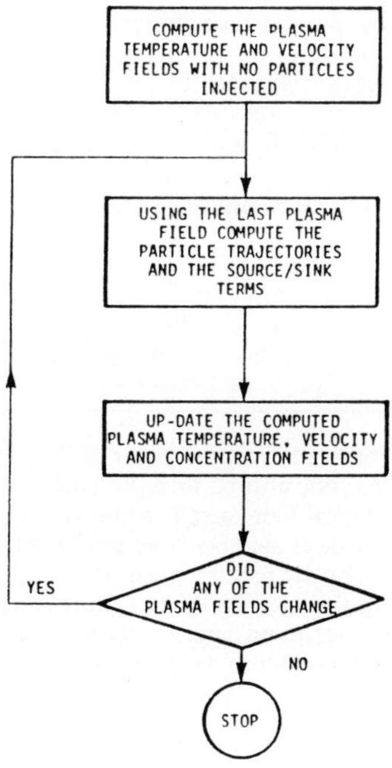

Figure 46: Flow chart of the computational method for high loading conditions [52].

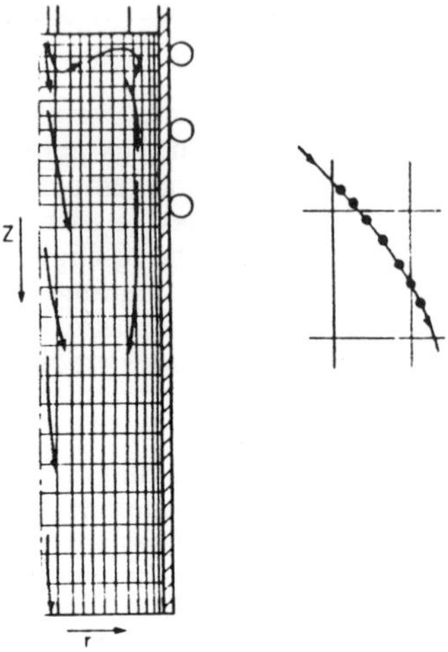

Figure 47: Computational domain in an RF plasma generator and typical cell [52].

5.2 Typical results for an r.f. plasma with copper particles

As an example of possible plasma-particle interaction effects in induction plasma modelling under heavy loading conditions, results reported by Proulx et al [55] will be given in this section. These were obtained for an inductively coupled plasma torch operated with argon at atmospheric pressure. Details of the torch dimensions and operating conditions are given in Table 4.

Table 4: Torch dimension and summary of the operating conditions

r_1=	1.7 (mm)	L_1	= 10.0 (mm)	Q_1	= 3.0 (slm)
r_2=	3.7 (mm)	L_2	= 74.0 (mm)	Q_2	= 3.0 (slm)
r_3=	8.8 (mm)	L_T	=250.0 (mm)	Q_3	=14.0 (slm)
R_o=	25.0(mm)	w	= 2.0 (mm)	P_o	= 3.0 (kW)
R_c=	33.0 (mm)			f	= 3.0 (MHz)

Copper powder with a mean particle diameter of 70 μm and a standard deviation of 30 μm is injected through the central tube into the coil region of the discharge. The physical properties of copper which were used in the calculations are summarized in Table 5. The choice of copper made it also possible to take into account the effect of the presence of copper vapor on the transport properties of the argon under plasma condition. For these, the data published by Mostaghimi and Pfender [57] were used, which showed that the presence of even small amounts of copper vapor can have a pronounced effect on the electrical conductivity of argon.

Table 5: Particle size distribution, feed rates and overall energy exchange between the powder and the plasma

Material: Copper powder, $\overline{d_p}$= 70 μm, $\overline{\sigma_s}$ = 30 μm, ρ_p = 8900 kg/m^3

T_m = 1356 K, H_m = 204.7 KJ/kg, c_{ps} = 0.425 kJ/kg
T_v = 2840 K, H_v = 4794.0 KJ/kg, c_{pv} = 0.480 kJ/kg

$\dot{m}_p$ (g/min)	$\dot{m}_v$ (g/min)	Q_p (W)	Q_v (W)	% of total energy absorbed
1.0	0.50	71.0	21.0	3.1
5.0	1.40	255.0	42.0	9.9
10.0	1.40	386.0	35.0	14.0
15.0	1.16	460.0	28.0	16.3
20.0	0.94	511.0	23.0	17.8

Dividing the Gaussian particle size distribution of the copper powder into, n_d, discrete fractions and their spatial distribution in the powder injection tube into, n_r, discrete positions, the problem simplifies to one of calculating n_d x n_r different possible trajectories. Assuming that the injection velocity of the particles are equal to that of the carrier gas velocity at the point of injection, 35 such individual trajectories could be calculated corresponding to values of n_d and n_r of 7 particle diameters and 5 injection points respectively. While these represent a first approximation they do not necessarily account for all possible tajectories (see section 4.3.2. C)

Results were reported [55] for different mass feed rates of the copper powder varying between 1.0 and 20.0 g/min. Figures 5-31 shows the isotherms, streamlines, and the concentration of cooper vapor in the torch for a feed rate of 5 g/min.

Because the trajectories of the particles in this case are very close to the axis of the torch, the plasma gas is significantly cooled down in this region (figure 49). On the other hand, the outer region of the plasma, where most of the power is dissipated, remains largely unaffected by the increase in the copper feed rate (figure 50). The result clearly demonstrates that although the overall loading ratio of the copper powder to plasma gas might be small (0.19 g copper/g argon), the local cooling effects are significant. This is a clear indication that the plasma-particle interaction effects could be locally very important under loading conditions for which it is generally accepted to neglect the changes in plasma temperature due to the presence of the powder.

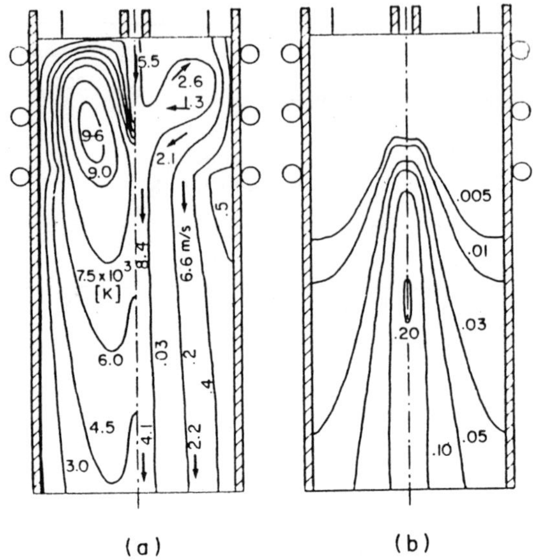

Figure 48: Isotherms and streamlines (a) and copper vapor isocontours (b) for copper injected in an argon plasma at feed rate of 5.0 g/min., after Proulx et al [55]

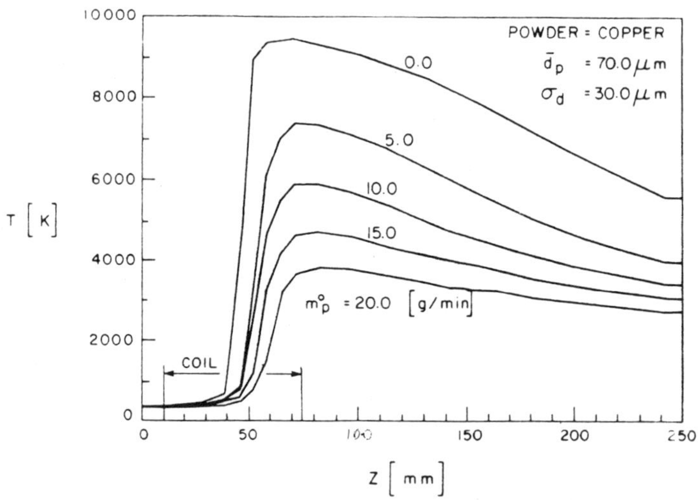

Figure 49: Effect of particle feed rate on the temperature profile along the axis of an induction plasma torch, after Proulx et al [55]

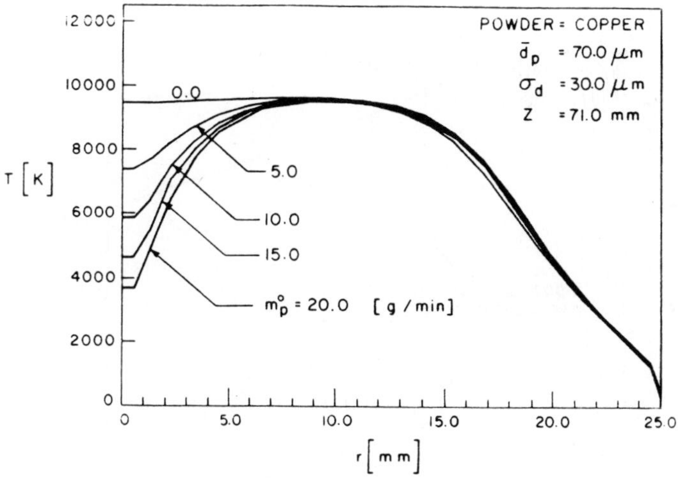

Figure 50: Effect of the powder feed rate on the radial temperature profile at z = 71.0 mm for an induction plasma torch, after Proulx et al [55]

It should be noted that momentum transfer between the gas and the particles is found to be negligible, and that the flow field is affected only through changes of the local plasma temperature. As the particles pass through the plasma, a portion of the powder evaporates and the vapor diffuses into the plasma medium (figure 48). The heat absorbed by the solid particles Q_p, and the superheat absorbed by the copper vapor to heat up to the plasma temperature Q_v, are given in Table 5. Q_v is generally much smaller than Q_p. The ratio Q_v/Q_p changes from 0.296 at 1 (g/min) feed rate to 0.045 at 20 (g/min) powder feed rate, and the total energy absorbed by the powder is between 3.1% and 17.8% of the input plasma power.

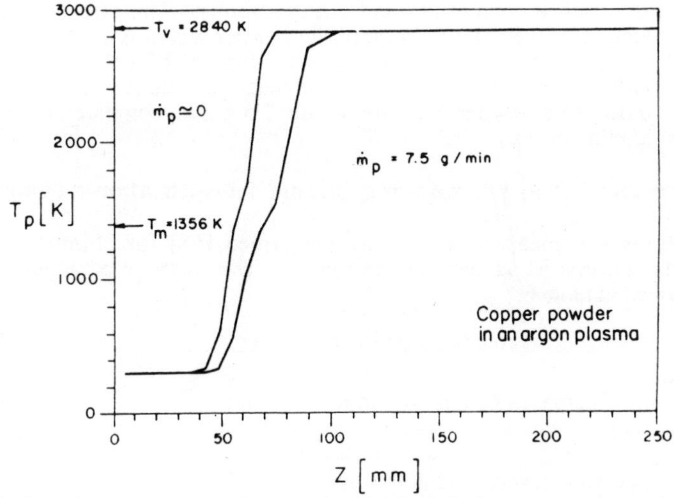

Figure 51: Particle temperature history along the axis of the discharge in presence and absence of particle loading effects after Proulx et al [55]

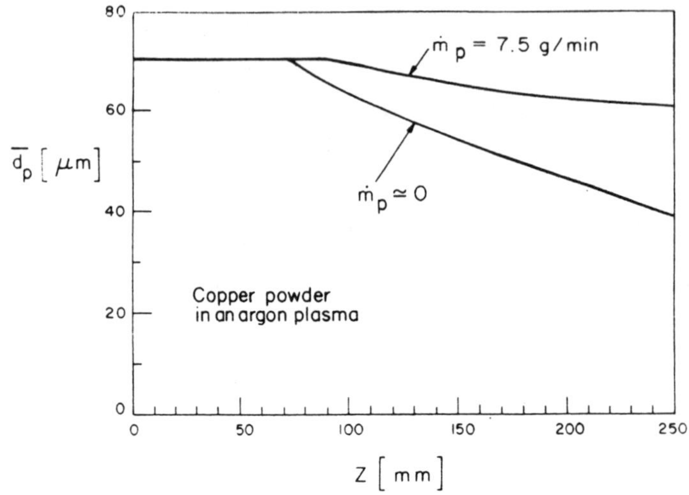

Figure 52: Mean particle diameter along the axis of the discharge in the presence and absence of copper loading effects after Proulx et al [55]

It should be pointed out that the plasma-particle interaction effect, through its influence on the flow and temperature fields in the discharge, could obviously have an important influence on the predicted particle processing efficiency . Figures 51 and 52 give typical particle temperature history profiles and average particle size, respectively in the presence and absence of loading effects.

It may be noted from figure 51 that the local cooling of the plasma due to the presence of the particles is responsible for the reduction of the heating rate of the particles as they are injected into the torch. This results in a downstream shift of the point at which the particles reach their melting and boiling points. The effects result in a corresponding reduction of the mass fraction of the powder which is evaporated by the time the powder exits from the discharge. As shown in figure 52 for the copper powder with an initial mean particle diameter of 70 μm, the mean particle size at the exit of the torch is close to 60 μm for a powder feed rate of only 7.5 g/min, compared to a value of 40 μm for the no-loading case.

5.3 Typical results for a d.c. spraying plasma jet with alumina particles

Calculations and measurements were performed [58] for alumina particles injected into a d.c. plasma jet as described in Fig. 41. The spraying parameters in this case are summarized in table 6.

Table 6. Spraying parameters

Nozzle i.d. 7 mm
Injector i.d. 1.8 mm, external injection r = 9 mm, z = 6 mm
32 slm Ar, 12 slm H_2
I = 600 A, V = 58 V, η = 52%
Al_2O_3 particles fused and crashed - 62 + 18 μm

The distributions of particle velocity and surface temperature were measured at a point situated on the jet centerline at 120 mm from the nozzle exit for different powder mass flowrates. Results given in Fig. 53 and 54 show that the velocity of the particulates is reduced by 11 % when the powder mass flowrate is increased from 0.03 to 2 kg/h while their surface temperature is reduced by about 14 %.

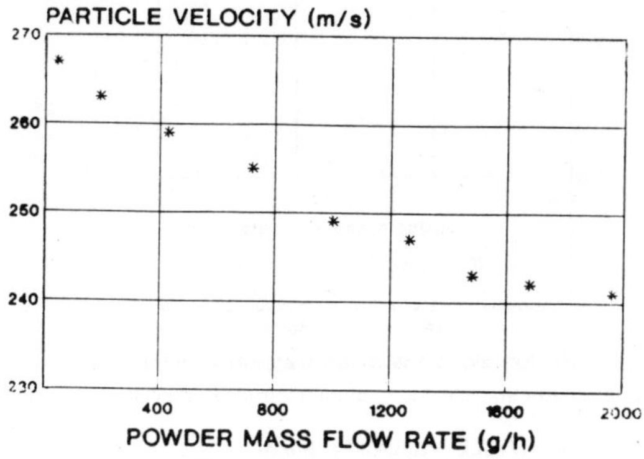

Figure 53: Al_2O_3 Particle velocity as function of the powder feed rate at 120 mm from the nozzle exit, after Vardelle et al [58].

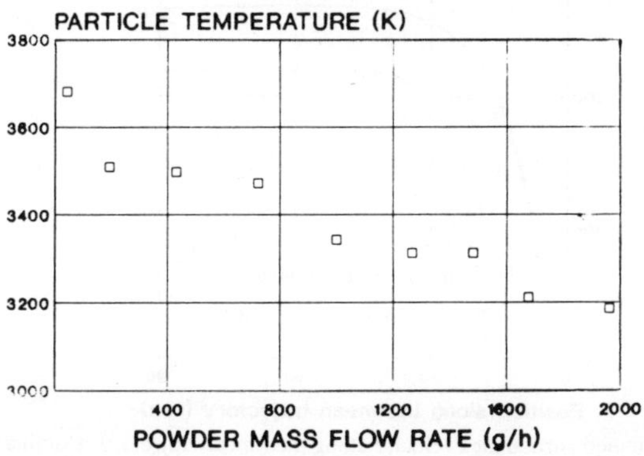

Figure 54: Al_2O_3 particle temperature as function of the powder feed rate at 120 mm from the nozzle exit, after Vardelle et al [58].

The evolution of the measured velocity of Al_2O_3 particulates along their "mean" trajectory is represented in Figure 55 for two powder feed rates (0.3 and 2 kg/h). The corresponding modeling results calculated for the same conditions are shown in Figure 56. From the comparison of the experimental and computed data, it is observed that for a low feeding rate, the mean calculated velocity is within 10 % of the experimental values. Significant differences are however observed when the powder mass feed rate is increased up to 2 kg/h. The particle loading effect is observed to be over predicted by the model (a reduction of 18% of particle velocity) compared with the measured values (11% reduction of particle velocity).

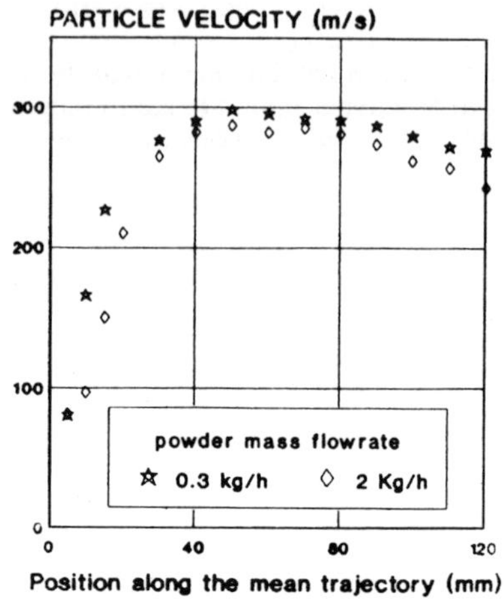

Figure 55: Measured particulates velocity along their mean trajectory (alumina powder)

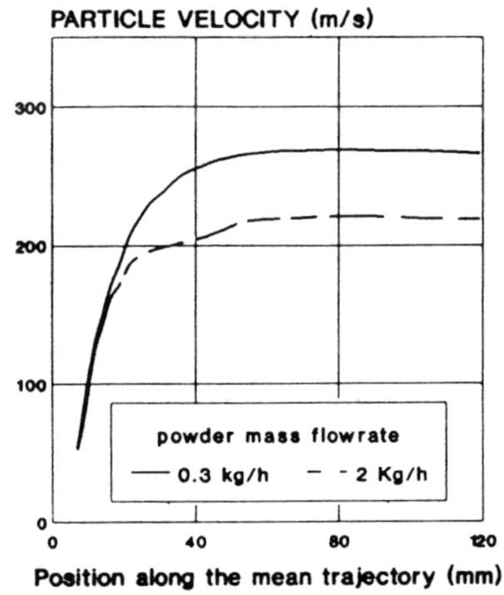

Figure 56: Computed particulates velocity along their mean trajectory alumina powder)

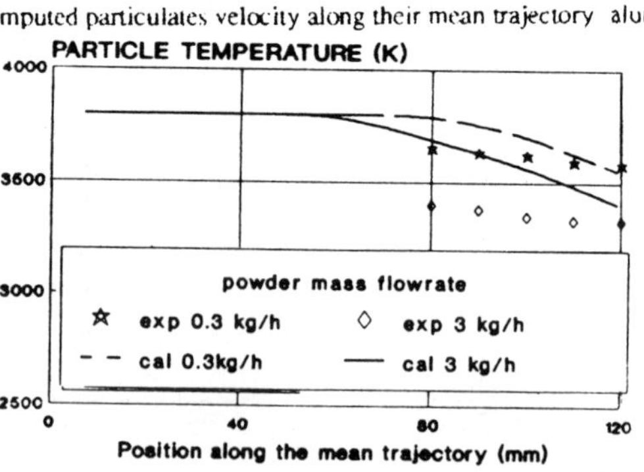

Figure 57: Computed and measured surface temperature of alumina particles along their mean trajectory, after Vardelle et al [58].

On the other hand, the loading effect on the particle surface temperature seems to be under estimated by the model compared to the experimental values as shown in Fig. 57. The discrepancy is within 8% in the plasma plume (particle temperature measurements are not feasible closer than 80 mm from the nozzle exit with our pyrometric device due to background radiation from the plasma). The effect seems to be related to the fact that the model tends to under estimate the expansion of the particulate stream in the plasma flow. Here again the remarks about the injection problem (see section 4.3.2-c) are of primary importance. Some alternatives in the computed injection conditions of the particles into the plasma jet might improve the models predictions.

6. Summary and conclusions

In conclusion it may be stated that the success of thermal processing of powders under plasma conditions depends to a large extent on our understanding of the basic phenomena involved in plasma-particle momentum and heat transfer. Of particular importance are the following phenomena which have been discussed in this chapter:

- Effects of the particle carrier gas on the flow and temperature fields in the plasma reactor.
- Effects of steep temperature gradients on the plasma-particle momentum and heat transfer.
- Effects of internal heat conduction within the particle.
- Effects of particle evaporation and non-continuum effects

Special attention has also been given to plasma-particle interaction under dense loading conditions which could have a significant influence on the particle processing efficiency and should be taken into account in the design of all powder feeding configuration in a plasma system. The way to represent the true trajectories of particles at the injector exit is still an involved problem. It requires more measurements of the particle velocities (along the axis and the radius of the powder injector) and their diameter, and the development of new models to account for the collisions of the particles with the injector wall which can have an important effect on the injection velocity distribution of the particles into the plasma.

References

[1] M. Vardelle , A. Vardelle , P.Fauchais, 1st Plasma Technik Symposium Wohlen Switzerland, 2, 167, (1988).
[2] M. Vardelle , A. Vardelle , P.Fauchais, Thermal Spray Technology (to be published).
[3] L. Gyenis, et al, 2nd Plasma Technik Symposium Lucern Switzerland, 1, 95 (1991) (Pub.).
[4] M. Vardelle, Thesis de doctorat, Univ. of Limoges, France, July (1988)
[5] F. Monerie-Moulin, F. Gitzhofer, P. Fauchais, M.Boulos., Journal of High Temperature Chemistry (to be published).
[6] W. Roman et al, In Thermal Spray Research and Applications, 49 (1991)
[7] Boulos et al, "Bubbles, Drops and Particles", Academic press (1978).
[8] R. Clift, J.R. Grace, J.E. Weber, "Bubbles, Drops and Particles", Academic press (1978).
[9] J.H. Masliyah, Ph.D. Thesis, Univ. of British Columbia, Vancouver (1978).
[10] E. Pfender, Y.C. Lee, Plasma Chem. and Plasma Process. 5, 211 (1985).

[11] N.N. Sayegh, W.H. Gauvin, AIChEJ, 25, 522 (1979).
[12] E. Pfender, Pure & Appl. Chem., 57, 1179 (1985).
[13] J.A. Lewis, W.H. Gauvin, AIChEJ, 19, 982 (1973).
[14] Y.C. Lee, K. Hsu, E. Pfender, 5th International Symposium on Plasma Chemistry, Edinburgh, U.K. 2, 795 (1981).
[15] E. Pfender, J. Plasma Chemistry and Plasma Processing, 9, 1675 (1989)
[16] Y.C. Lee, Master Thesis, Univ. of Minneapolis, MN, USA (1982)
[17] A. Vardelle, Thesis de doctorat, Univ. of Limoges, France, July (1988).
[18] B.Pateyron, M.F. Elchinger, G. Delluc, P. Fauchais, Thermodynamic and t ransport properties of Ar-H_2 and Ar-He plasma gases used for spraying, accepted in Plasma Chemistry and Plasma Processing, September (1992)
[19] E.Bourdin, P.Fauchais, M.I. Boulos, Int. J. Heat and Mass Transfer, 26, 567 (1983).
[20] J.K. Fizdon, Int. J. Heat Mass Transfer, 22, 749 (1979).
[21] X.I.Chen, In Heat Transfer Science and Technology (ed. by Wang B.X.) pp. 748-755., Hemisphere, Washington (1987)
[22] X.I.Chen, Pure and Applied Chemistry, 60, 651 (1988).
[23] X.IChen and E. Pfender, J. Plasma Chem. and Plasma Processing, 3, 97 (1983).
[24] Ibid, J. Plasma Chem. and Plasma Processing, 3, 351 (1983).
[25] Ibid, J. Plasma Chem. and Plasma Processing, 2, 293 (1982).
[26] M. Vardelle, A. Vardelle, P.Fauchais, A.Denoirjean, 12th Int. Thermal Spraying Conf., (Pub.), The Welding Institute, London, 1, E 49 (1989).
[27] Y.C.Lee, Y.P. Chyou, E.Pfender, Plasma Chemistry and Plasma Processing, 5, (4) 391 (1985).
[28] T. Yoshida, K.Akashi, J. Appli. Phys. 48, (6) 2252 (1977)
[29] W.E.Ranz, W.R. Marshall, Chem. Eng. Prog. 48 , (3) 141 (1952).
[30] Ibid, Chem. Eng. Prog. 48, (4), 173 (1952).
[31] X.I. Chen and E. Pfender, Plasma Chem. Plasma Proc. 2, 185 (1982).
[32] J.L. Engelke, Internal Report, Stanford Research Institute, Ca, USA (1962).
[33] C. Borgianni, M. Capitelli, F. Cramarossa, L.Triolo, L.Molinari, Combust. Flame, 13, 181 (1969).
[34] A. Essoltani, P.Proulx, M.I. Boulos, A.Gleizes, J. Analytical Atomic Spectrometry 5, 543 (1990).
[35] Ibid, Effect of the presence of metallic vapors on volumetric radiation losses under plasma conditions, to be published.
[36] Ibid, Heat transfer in thermal plasma processing ASME-HTD 161, 145 (1991).
[37] P. Humbert, Thèse de l'Unviersité de Paris VI, ENSCP, rue P et M Curie, Oct. (1991).
[38] M.I. Boulos and W.H. Gauvin, CJChEJ, 52, 355 (1974).
[39] D. Bhattacharyya and W.H. Gauvin, AIChEJ, 48, 879 (1975).
[40] T. Yoshida and J. Akashi, Appl. Phys., 48, 2252 (1977).
[41] M.I. Boulos, IEEE Trans. Pl. Sc., 4, 93 (1978).
[42] F.J. Harvey and T.N. Meyer, Metallurgical Trans., 9B, 615 (1978).
[43] J.K. Fizsdon, Int. J. Heat & Mass Transfer, 22, 749 (1979).
[44] J. Lesinski, B. Mizera-Lesinska, J.C. Fanton and M.I. Boulos, AIChEJ, 27, 358 (1981).
[45] Y.C.Lee, "Trajectories and heating of particles injected into a thermal plasma", M. Sc. Thesis, U. of Minn. (1982), USA.
[46] M. Vardell, A.Vardelle, P.Fauchais and M.I.Boulos, AIChEJ, 29, 236 (1983).

[47] M. Vardelle, A.Vardelle, P. Fauchais and M.I. Boulos, AIChEJ, 34, 567 (1988).
[48] P. Fauchais, A. Vardelle, M. Vardelle, F. Monnerie, A. Denoirjean, A. Grimard Thermal Spray Contings: Properties, Processes ans Applications, p 205 (Pub.) ASM Int, FL, USA (1992).
[49] P. Fauchais, J.D. Condert, M. Vardelle, Diagnostics in Thermal Plasma Processing, in Plasma Diagnostics, p. 349, (Pub.) Academic Press (1989), Boston, USA.
[50] P. Fauchais, J.F. Condert, M. Vardelle, A. Vardelle, A. Denoirjean, Diagnostics of Thermal spraying plasma jets, accepted in Journal of Thermal Spraying, Oct. (1992)
[51] M.I. Boulos, IEEE, Trans. Pl. Sc., PS-6, 93 (1978).
[52] P. Proulx, "Interaction plasma-particules dans la modélisation des plasmas à induction", mémoire de M. Sc. A., Université de Sherbrooke, avril (1984).
[53] P. Proulx, J. Mostaghimi and M.I. Boulos, 6th International Symposium on Plasma Chemistry, Montreal, July (1983), 1, p. 59 (1983).
[54] J. Mostaghimi, P. Proulx, M.I. Boulos and R.M. Barnes, Spectro-chemical Acta, 40B, 153 (1985).
[55] P. Proulx, J. Mostaghimi and M.I. Boulos, Int. J. Heat & Mass Transfer, 28, 1327 (1985).
[56] C.T. Crowe, M.P. Sharma, D.E. Stock, J. Fluids Eng. 99, 325 (1977).
[57] J. Mostaghimi, E. Pfender, Plasma Chemistry and Plasma Processing, 4, 199 (1984).
[58] A. Vardelle, M. Vardelle, P. Fauchais, P. Proulx, M.I. Boulos, Thermal Spray: Int. Advances in Coatings Technology, p.543 (Pub.) ASM Int. FL, USA (1992).

Chapter 2

DIAGNOSTICS OF THERMAL SPRAYING PLASMA JETS

P. Fauchais, J.F. Coudert, M. Vardelle, A. Vardelle, A. Denoirjean
Laboratoire de Céramiques Nouvelles
URA 320 CNRS - University of Limoges - FRANCE

ABSTRACT

D.C. thermal plasma jets are strongly affected on the one hand by the arc root fluctuations at the anode, resulting in a type of pulsed flow and enhanced turbulence, and on the other hand by the entrainment of surrounding cold gas in the plasma jet. These phenomena and the resulting temperature distributions have been studied using a wide range of diagnostic techniques including fast cameras, laser doppler anemometry (LDA), coherent anti-Stokes Raman spectroscopy (CARS), Rayleigh scattering, emission spectroscopy, Schlieren photography, enthalpy probes and sampling probes. The information given by these techniques is evaluated and compared. The effect of the arc fluctuations on the spectroscopic measurements is emphasized and the possibility of using these fluctuations to determine informations on the arc behaviour and the axial velocity of the jet is presented.

Optimization of plasma processing of solid particules requires information about their size and surface temperature, as well as number flux and velocity distributions at various locations in the flow field. The different statistical techniques of in-flight measurements are discussed together with their limitations. A method to determine the temperature and species density of the vapor cloud or comet travelling with each particule in flight is then presented. However, such statistical measurements present ambiguities in their interpretation, which can be adressed only by additional measurements to determine the velocity, diameter, and surface temperature of a single particule in flight. Moreover, information on single particules is required in order to understand the coating properties, which depend strongly on the way the particules flatten and solidify upon impact. A method to obtain data related to a single particule in flight and to follow the temperature evolution of the corresponding splat upon cooling is presented. The paper concludes with the description of the experimental techniques to follow the temperature evolution of the successive layers and passes. This is important because temperature distribution within coating and substrate controls the adhesion and cohesion of coatings as well as their residual stress.

1. INTRODUCTION

D.C. plasma spraying is a fast growing technology which has diffused progressively from aeronautic and nuclear industries to other industries. This is due to the increasing need for thick coatings with progressively more sophisticated properties and for better control of coating quality and reproducibility. These results have been possible thanks to a better understanding of the relevant phenomena resulting from the development of diagnostic techniques that will be summarized in this paper.

The thermomechanical properties of coatings as well as their adhesion and cohesion depend /1 to 5/ on :

a/ the molten state, velocity, size, and density of particules upon impact on the substrate. These parameters are controlled on the one hand by the heat and momentum transfers between the plasma jet and the particules and on the other hand by particule morphologies, size, and injection velocity distributions.

b/ the chemical reactions of the particules with the surrounding atmosphere, especially the entrained air, as well as their possible decomposition or bursting upon their penetration into the plasma jet.

c/ the particule flattening and cooling is linked to the conditions recalled in a/, but also to the nature, roughness, and temperature of the substrate or the previously deposited layers. These parameters control the cohesion and adhesion of the coating as well as the quenching stresses. Moreover, the heat transfer between the plasma plume, the successive beads and passes, and the substrate induces temperature gradients during spraying which vary during cooling. These variations control the residual stresses and cracking within the coating.

The heat and momentum transfer between the plasma jet and the particules, depend among other parameters, on the temperature and velocity distributions of the plasma jet. Therefore modelling and prediction of these quantities is mandatory. However, many problems are still pending for such measurements. In general, this is due to the continuous fluctuations of the plasma jet /6/ with some sort of pulsed flow that is related to the surrounding gas entrainment mechanism. It is an engulfment type process /7/ that plays an important role with regard to chemical reactions in flight and cooling of the plasma jet. The temperatures needed for the calculations are the electron (T_e) and heavy species (T_h) temperatures which are not directly accessible. For velocity no method seems to give reliable results /8,9/ at the present time. To calibrate the models, it is also necessary to measure the velocity, number density, diameter, and surface temperature of the particules in flight. However, for d.c. plasma jets only distributions are measured ; and the interpretation of the results, especially for surface temperature, is not straight forward /10,11/.

A basic understanding of plasma coating formation requires the knowledge of the thermal history of the droplets during their flattening and cooling process. These mechanisms have received much less attention than those related to particules in flight /12/. Although many micrographs of single particules collected during spraying have been published, their interpretation is still questionable due to the lack of information about the velocity, diameter, and surface temperature of the particules upon impact as well as the cooling rates of the splats. The first known measurements of droplets cooling during and after flattening have been available only very recently. The cooling of the splats from more than 35OO K to less than 2000 K takes place in a few tens of microseconds and seems to depend not only on the substrate or the nature of the previously deposited layers but also on their temperature. The material deposited in a single pass of the torch (resulting in a single bead with a roughly Gaussian shape, see section 4) cools down in times of the order of milliseconds at temperatures below 2000 K depending upon factors such as the powder feed rate and the velocity of the torch relative to the substrate. The cooling rate is also affected by the relative motion torch to substrate, the time and amount of overlap between successive beads (resulting in passes), and the possible use of auxilliary cooling devices.

Temperature gradients within the substrate and the coating, cause residual stresses within the coating and substrate that can induce cracking, especially in ceramic deposits.

2. PLASMA JET MEASUREMENTS

2.1. General remarks

Before describing the experimental measurements, it is important to recall briefly how the plasma jet is generated. The plasma column expansion from a tiny molten spot (a few tenths of mm^2) at the cathode tip, is strongly affected by the cold plasma gas flow close to the cathode tip, as well as the nature and flowrate of the gas and the arc current /13 to 16/. This flow depends on the cathode cone angle and diameter, the arc chamber and gas injector designs, as well as the nozzle diameter and design.

When the cold boundary layer close to the anode-nozzle wall is hot enough (and this is correlated to the plasma column expansion), the arc strikes at the anode wall by one or a few tiny plasma columns which are continuously fluctuating under the action of the drag force and the j x B forces /13/. These fluctuations continuously modify the length of the electrically conducting plasma column. Thus, the voltage presents a saw tooth shape, with frequencies varying between one and a few tens of kHz /15,17/. These fluctuations of the plasma column length (linked to the gas aerodynamics in the nozzle) coupled to the negative resistance of the arc, react with the power supply, which brings in its own fluctuations (at frequencies below 600 Hz in general). All these phenomena induce pulsating plasma jets /18/. These pulsations, together with the large velocity difference between the plasma jet and the surrounding atmosphere, cause large scale eddies and the onset of turbulent flow /7/.

To summarize, the plasma jets which are to be analysed :

- are continuously fluctuating (equivalent to some sort of pulsing flow) with a certain correlation between electric and aerodynamic fluctuations,

- are highly turbulent in their fringes and in their plume, with non isotropic velocity fluctuations /17/,

- have an entrainment mechanism of engulfment type /19/ with cold surrounding gas "bubbles" (higher density and thus inertia than their high temperature surroundings) entrained in the hot plasma jet. These bubbles do not mix with the plasma, at least initially.

- present high radial temperature gradients (up to 4000 K/mm) /8/.

2.2. Arc fluctuations and turbulences

2.2.1. Video cameras

Computerized video-image processing and process-control software make it possible to automatically extract information from video-images. For example in Control Vision's laser enhanced stroboscopic video system /19/, an electronic image of the process is captured using momentarily intense light from a strobe-illumination unit rather than using the light from the process itself. The system provides real time pictorial information on the dynamic behaviour of the plasma jet or of particules (in solid or molten state) moving at high speeds within the most intense plasma jets /18/.

For the plasma jet fluctuations, it is also possible to use video-cameras with shutter speeds up to 10^{-4} - 10^{-5} s.

Such measurements (performed with strobe-illumination /21/, or with a simple video camera) have shown that the plasma jet, which with the naked eye looks stable and axisymmetric, is, in fact, continuously fluctuating in length (up to 50-60%) and in position. The extremity of the bright core moves radially and axially in a random manner. For a given nozzle diameter, these fluctuations seem to depend on the flow close to the cathode tip, the gas nature and flowrate, and the arc current. However, to obtain more precise information (such camera give ten to sixty frames/s) other measurements are necessary.

2.2.2. Spectral analysis

Spectral analysis is a non intrusive diagnostic technique capable of providing a large amount of information about the operating conditions of a plasma torch /17/. Analysis of pressure fluctuation, for example through acoustical spectral measurements with an omnidirectional microphone, provides information about the flow structure due to the arc root motion, especially in the fringes of the plasma column. Further, by spectrally analyzing the signals from a photomultiplier or a photodiode focussed on the axis of the plasma jet at a given location, one can gain insight into the temperature fluctuations of the jet and how they correlate with the arc root motion. The voltage fluctuations are due to the axial motion of the arc column and/or of the anode arc root. Coupling with the power source response can be deduced from the simultaneous study of arc current fluctuations. The first measurements presented by Pfender et al /17/ seem to show that the relationship between internal and external perturbations is rather weak. The main parameters influencing the fluctuations seem to be the same as those already underlined for the video camera measurements, with a more pronounced effect of the flowrate. However, an appropriate signal treatment with adapted filters /22/ before cross correlation seems to give better correlations. These measurements could also probably give information about electrode erosion, but considerable additional work is needed to better understand the involved phenomena.

2.2.3. Shadowgraphy

Shadowgraphs of plasma jets /23,24/ reveal the abrupt development of turbulence from a well behaved laminar flow near the nozzle exit. They reveal typical shear layer instabilities close to the nozzle exit, characterized by the formation of vortex rings around the plasma jet. Farther downstream, the vortex structure breaks down over a relatively short distance, resulting in large-scale turbulence eddies which rapidly penetrate into the core of the plasma jet. The location of the transition zone /24/ is thought to be driven by the rapid cooling of the jet and the resulting increase of the Reynolds number (from about two orders of magnitude at the torch exit to four orders of magnitude in the transition zone). Here also the arc current and the gas flowrate play an important role /24/. However, the respective influences of the pulsations of the plasma jet and of the large velocity difference between the plasma jet and the surrounding atmosphere in the development of the turbulences are not yet understood.

2.2.4. Turbulence measurements

Turbulence intensity measurements have been performed /17,24/ by using Laser Doppler Anemometry of small alumina particules (d<3 mm). They were injected from a specially designed fluidized bed /25/, inside the nozzle downstream of the arc root. The turbulence was defined as the ratio of the standard deviation of the individual velocity measurements for a given data point divided by the maximum centerline mean velocity for a given axial distance downstream. As for the shadowgraphs, the results show the rapid increase of the axial velocity fluctuations at the edge of jet due to the mixing and entrainment of external air.

By injecting the small alumina particules around the plasma jet the velocity of the entrained air was measured along the centerline of the plasma /24/. Such measurements represented in fig. 1 show that, with the given conditions, for $z \leq 30$ mm the fluid from the plasma jet has a much higher mean velocity than the entrained gas which has reached the center line at a distance of about 8-10 mm. Entrained air does not mix well with the plasma gas initially, but instead, travels with a much slower mean velocity than the plasma, thus forcing the plasma gas to flow around it. Such measurements are confirmed by CARS (see section 2.3.3.). This seems to indicate that the more classical belief of entrainment and mixing by stochastic small-scale eddies is not a realistic picture of the transport mechanisms in turbulent plasma jets.

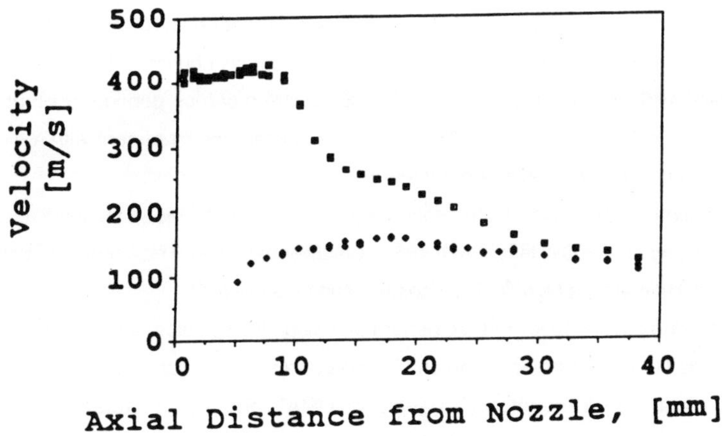

Fig. 1 : Comparison of plasma jet and entrained fluid velocity on jet centerline (Ar plasma, 23.6 slm, 450 A, 24V) after Pfender et al./24/.

2.3. Temperature, velocity, and concentration measurements

2.3.1. Instant or mean values

The techniques used are mainly emission spectroscopy, laser scattering and enthalpy probes. Enthalpy probes give Favre averaged values and cannot follow the plasma jet fluctuations and/or turbulence. Local measurements (according to the size of the probe) can be performed.

The measurement, by laser anemometry, of the velocity of small particles injected in the plasma jet can give the time averaged flow velocity provided they have travelled a minimum distance in the jet before the measurement point in order to overcome the Knudsen effect /17,24/. Such measurements also allow to determine the velocity dispersion due to the different treatments undergone by the particules according to the dimensions of the measurement volume and the size of the particules which is not necessarily small compared to the turbulence eddies. For example Spores /26/ indicates that up to 10% of the measured dispersion, often called turbulence, is associated with the size distribution of the seed particules.

For CARS experiments, the entire CARS spectrum can be generated simultaneously on the time scale of the laser pulse /23/. It is thus an instant measurement with a high spatial resolution when using crossed beam geometry.

Rayleigh scattering measurements are performed using either a pulsed Yag laser with boxcar averaging /27/ or CW lasers with lock-in amplifier detection. Thus, time averaged results are obtained. Here also, a high spatial resolution can be achieved.

Emission spectroscopy can be performed using either photomultipliers with a very fast response time (10 ns can be achieved and large fluctuations correlated to those of the plasma jet can be observed for the intensity signals) or OMA* giving time averaged signals. Such measurements are, of course, along the line of sight /8/ and Abel's inversion has to be performed to obtain local measurements.

As the relationship between the line intensity and temperature is not at all linear, a time averaged intensity does not permit direct determination of temperature from time-averaged measurements. The signal has to be processed to extract the stationary component of temperature /28/, which may be quite different from the temperature deduced from the time averaged signal, as illustrated in fig. 2. In this figure, the error bars marked "gaus" corresponds to the temperature fluctuations due to the random attachment of the arc root in the nozzle (described by a Gaussian noise) while T_{cos} correspond to the power source fluctuations.

* *OMA : Optical Mutlichannel Analyser.*

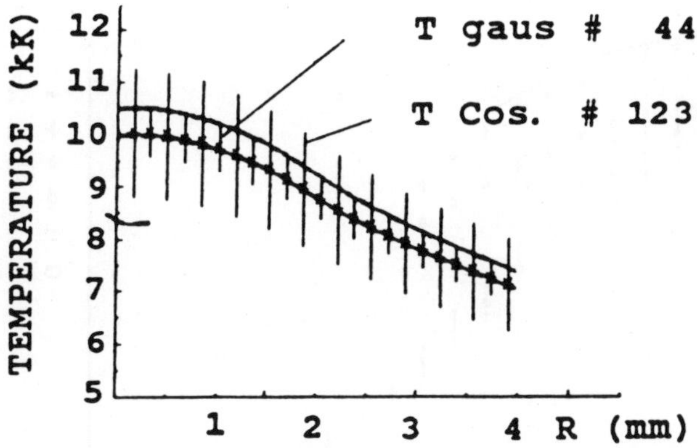

Fig. 2 : Comparison of the radial temperature distribution deduced from the time averaged intensity (—) and the stationnary component of the temperature obtained after signal treatment (static rectifier, P=23.2 kW, 290 A, 50 slm N_2, anode i.d. 7 mm) after /28/.

2.3.2. Temperature measurements

a/ Emission spectroscopy

Such measurements give the population of the emitting level and assumptions must be made to determine a temperature from this population /8/. Thus, depending on such assumptions, different temperatures can be measured :

. Rotational temperature, from 4000 to 11000 K (in general from N_2^+ spectra) with either a few percent of nitrogen introduced in the plasma jet or using that of the entrained air. This temperature can generally be assumed to be that of the heavy species T_h, due to the very short relaxation time between translation and rotation (~10^{-8} s) /8/.

. Excitation or population temperatures from atomic lines which values are between those of heavy species and electrons (7000 < T < 15000 K).

. Temperatures deduced from the continuum nearer to T_e than to T_h (8000 < T < 15000 K).

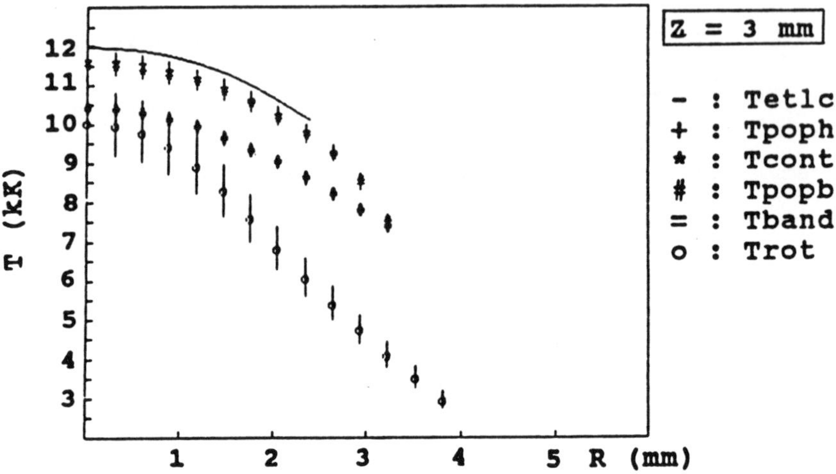

Fig. 3 : Temperature distributions measured 3 mm downstream of the nozzle exit of a D.C. plasma jet (cylindrical nozzle 6 mm in diameter, 250 A, 82 V, thermal efficiency 71%, 40 slm N_2, p_{ch}=33 Pa). T_{poph} NI 532.8 nm ; T_{popb} NI 746.8 nm ; T_{band} : band head of N_2^+ 1^-, 0-0 ; T_{rot} : rotational spectra of N_2^+ 1^-, 0-0 ; T_{cont} : λ=531 nm.

In this figure, the error bars for T_{rot} corresponds to :

$$\delta T = \sqrt{2Em/\frac{\partial^2 Eq}{\partial T^2}}$$

where E_m is the minimum of a quantity which allows to determine T_{rot} /16/ and is related to the signal to noise ratio, while $\partial^2 Eq/\partial T^2$ is related to the sensitivity of the method.

These different temperatures are not necessarily in equilibrium among themselves. This is illustrated in fig. 3 for a d.c. nitrogen plasma jet flowing in a controlled atmosphere chamber filled with nitrogen at a pressure of 33 kPa /16/. Of course, in such pressure conditions the non-equilibrium effects are somewhat enhanced, but the trends are the same as those at atmospheric pressure. The temperatures deduced from atomic excited states (T_{poph}, T_{popb}), from continuum (T_{cont}), from the band head (T_{band}) are higher than those obtained from the rotational spectra, especially in the fringes of the jet where the diffusion of the electrons is important /29/.

Similar results were obtained with argon plasmas /30/. This clearly indicates that measurements performed with atomic lines have to be considered as trend indicators that give only relative information, for example, to compare torches or working conditions.

For example, fig. 4 illustrates the cooling of an Ar-H_2 plasma jet at atmospheric pressure when flowing successively in argon (ionization at 15000 K), nitrogen (dissociation at 7000 K) and air (oxygen dissociation at 3500 K). It can be seen also with these time averaged measurements that the jets are rather symmetrical as when looking at them with the naked eye.

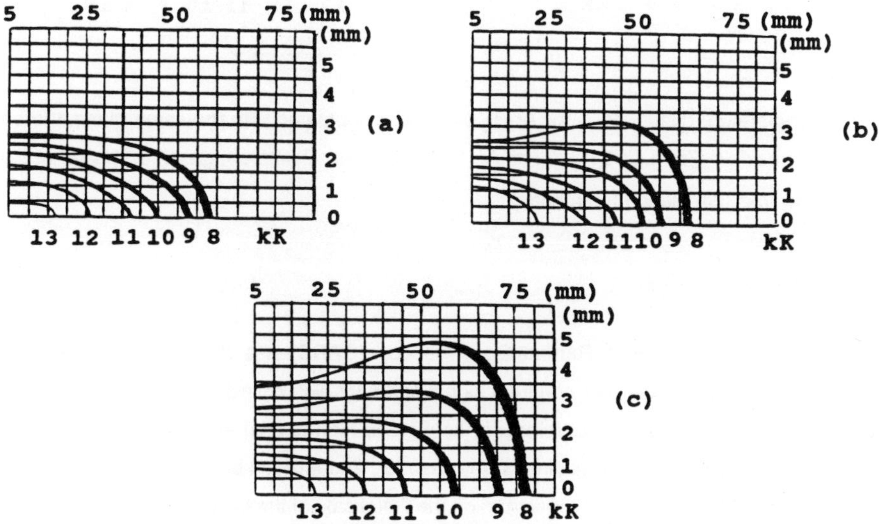

Fig. 4 : Temperature distributions for an Ar-H_2 plasma jet at atmospheric pressure (P=35 kW, 81 slm Ar, 9 slm H_2) flowing in air (a), nitrogen (b), argon (c) /31/.

Such measurements /31/ also confirm the influence of the plasma gas injector design on the turbulent mixing with the surrounding air, as illustrated in fig. 5. It can be seen that when injecting the plasma gas radially close to the cathode tip, the length of the plasma jet is drastically reduced compared to that obtained when injecting the gas along the cathode. In both cases, the voltages are the same within 2%, and the thermal efficiencies differ by less than 3%. These time averaged measurements demonstrate the strong coupling of the turbulent mixing with the "pulsed flow".

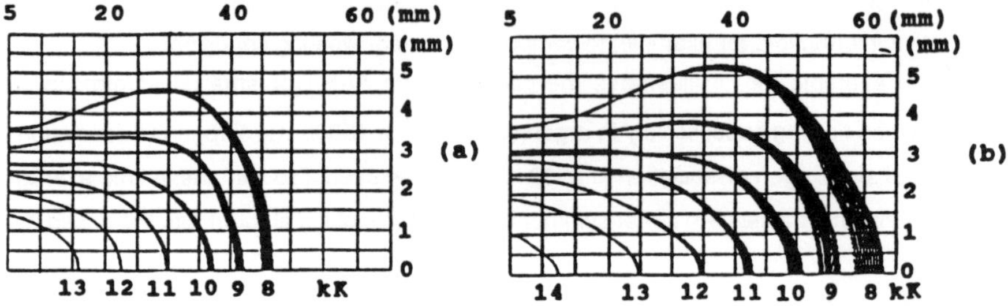

Fig. 5 : Temperature distributions for Ar-H_2 plasma jets obtained with the same plasma torch (cylindrical nozzle d=7mm, 500 A, 30 slm Ar, 12 slm H_2) with 2 different gas injectors (a) radial injection, (b) injection along the cathode.

b/ Laser scattering

Coherent Anti-Stokes Raman Spectroscopy (CARS) has the advantage of high conversion efficiency, a laser-like coherent signal beam for high collection efficiency, excellent fluorescence and luminosity discrimination, as well as high spatial and temporal resolution /32,33,34/. It has been used for plasma spray jets /23/, and a broad band approach has been used to generate the entire CARS spectrum on the time scale of the laser pulse. Temperatures were derived from the spectral distribution of the CARS signal using a quick-fit technique based on the full-width at half maximum of the spectral signal. The temperature determination was calibrated against a Pt, Pt-Rd thermocouple. The nitrogen molecular spectra have been used to determine the temperature of surrounding air pumped by an argon plasma jet /24/ as shown in fig. 6.

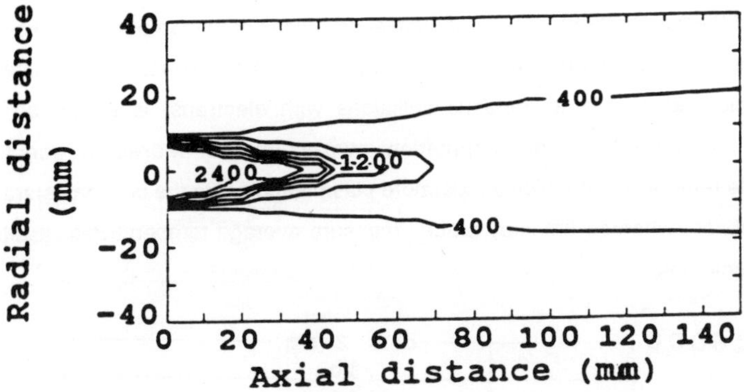

Fig. 6 : Contour plots of nitrogen temperature for an argon plasma jet in air (250 A, 9.8 slm Ar, 10 kW) after Fincke et al./23,24/.

Integrated Rayleigh scattering measurements have been performed in the fringes of argon D.C. arcs or jets flowing in argon, using either chopped CW lasers /35,36/ or pulsed Nd : YAG lasers /27/. The light scattered through an angle of 90° was focused onto the entrance slit of a monochromator and detected using a photomultiplier. Such measurements give the density of the neutral species, provided that the scattering cross sections are known. At atmospheric pressure, the density of such particles becomes very low for temperatures higher than 8000 K, which is about the limit of the method. Such measurements, which have to be calibrated with gases of known temperatures, also show a strong discrepancy in the jet fringes when comparing the temperatures measured by emission spectroscopy from atomic lines and those deduced from Rayleigh scattering (see for example /27/). Problems related to measurements in gas mixtures, such as Ar-H_2-Air (air pumped in the plasma plume), must also be addressed.

c/ Enthalpy probes

New computer controlled systems of small size (d < 3 mm), inducing low perturbations of the flow, are now available /17,24/ to determine temperatures between 11000 and 2000 K in argon plasma jets and somewhat lower temperatures in Ar-H_2 jets (maximum heat fluxes of 10^6 W/m^2). Fig. 7 from /24/ shows some typical results obtained in an argon D.C. plasma jet and compared to population temperatures determined by emission spectroscopy. The comparison of the two methods shows substantial deviations of the temperatures, especially in the jet fringes. With the fast diffusion of electrons in the cold regions, the population of the excited atomic states, due to inelastic collisions with electrons, is enhanced and the temperature deduced from this population over-estimated. Moreover spectrometric measurements refer only to the high temperature plasma and not to the cold gas entrained into the plasma plume, whereas enthalpy probes measure average temperatures resulting from both hot and cold gas.

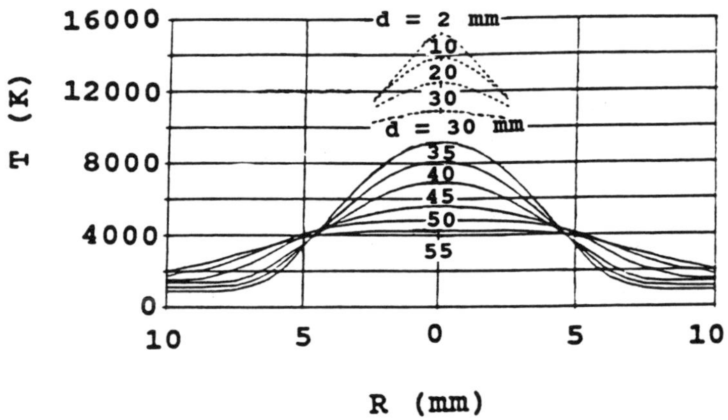

(... from atomic emission spectroscopy, — from enthalpy probes)

Fig. 7 : Temperature profiles of an argon plasma jet in argon atmosphere (400 A, 100 scfh Ar, 47 scfh He, p=10^5 Pa, d=distance from nozzle) /24/.

2.3.3. Concentration measurements

CARS has been used in D.C. argon plasma jets to probe the average concentration of nitrogen, from the surrounding air, entrained in the plasma jet /24,37/. Concentration was derived from the strength of the CARS signal. The measurement was calibrated by varying the concentration of air premixed in a cold argon jet. Fig. 8 shows the resulting contour plots of nitrogen concentration for a 10 kW argon torch power. The rapid entrainment of air into the core flow of the jet, coincident with the onset of turbulence and jet break-up, is evident in the concentration contours. Such a transition region is not as evident in the associated temperature field (see fig. 6).

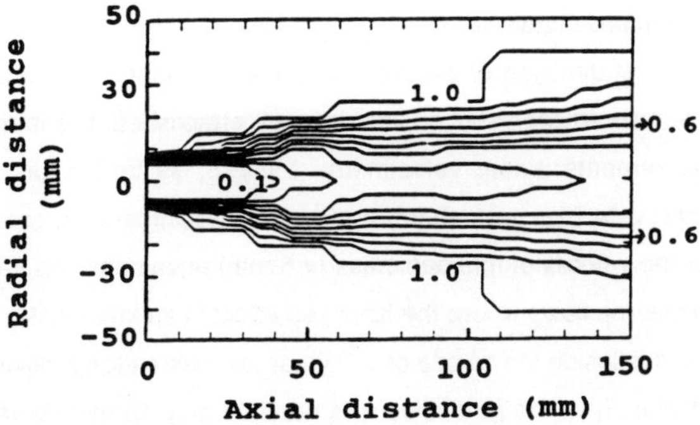

Fig. 8 : Contour plots of nitrogen concentration for a 10kW argon torch power (anode i.d. : 9.52 mm).

Experimentally, argon concentrations were derived from samples taken with an enthalpy probe /38,17/. The results, which were only available from z=20 mm and farther downstream, show a large drop in centerline concentration between the 25 and 35 mm profiles, indicating the end of the potential core. In good agreement with the CARS and LDA measurements (see sections 2.3.2.b and 2.1.4), more than 50% air was observed for distances greater than 30 mm. The air pumping increases with the gas flowrate. When the arc current increases, the air pumping occurs at longer distances due to the higher viscosity of the plasma jet. However, as shown by the spectroscopic measurements, when the arc column diameter becomes close to that of the nozzle, the length of the jet tends to a limit. This is illustrated in fig. 9, where it can be seen that the radius increases from 35 to 50 kW while the length remains almost constant.

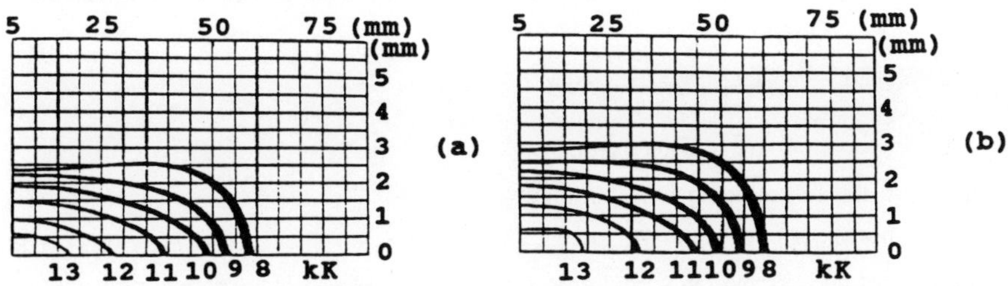

Fig. 9 : Temperature distributions for $Ar-H_2$ plasma jets : 75.6 slm Ar, 14.4 slm H_2, nozzle diameter 6 mm : a) 35 kW, b) 50 kW.

2.3.4. Velocity measurements

If temperature measurement methods are now well established, this is not the case for the flow velocity measurements. Laser velocimetry (L.V.) /8, 39 to 41/ has been used to measure solid particules velocities in a thermal plasma environment. In most gases, it is generally assumed that the velocity of fine particules (< 5 mm) equals the gas velocity, but this may not be the case in plasma flows where the Knudsen effect is enhanced /42/. For example, when injecting the particules inside the nozzle of a D.C. argon plasma torch downstream of the arc root, the maximum velocity of the particules was reached only 15 mm downstream of the nozzle exit /43/.

Velocity of argon plasma jets has also been determined by observing the Doppler shift of fluorescence lines /44,45/, but the obtained values seem to be overestimated. Pitot probes have been successfully used to determine a Favre averaged velocity at distances from the nozzle exit greater than 30 mm for argon D.C. plasma jets /46/.

Measurements based on either natural or imposed /47/ fluctuations of the jet have been developed. In the laboratory, the random attachment of the arc root in the nozzle has been used. It induces variations of the volumetric energy density in the plasma, and these homogeneous regions are entrained by the flow. The light intensity fluctuations sampled at two different points along the jet axis are correlated after a proper signal treatment : elimination of the noise from the first signal followed by a filtering of the second signal with a filter calculated with the help of the filtered first signal. In this case, a sharp cross-correlation function between the two filtered signals could be determined, permitting a measurement, with a reasonnable accuracy, of the time shift between the two signals and thus the axial velocity of the flow.

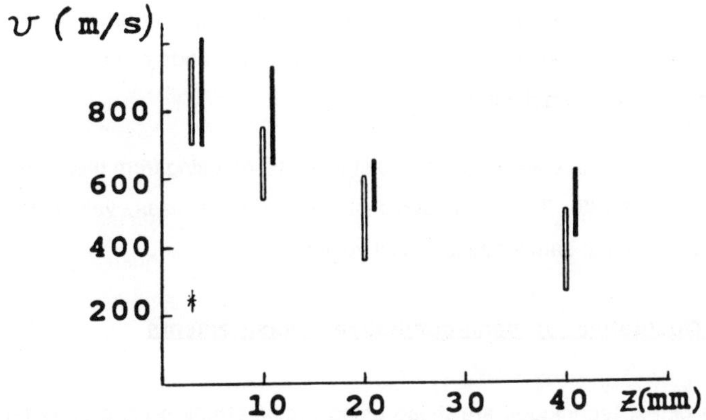

Fig. 10 : Evolution along the plasma jet axis of the velocity of the flow for three hydrogen percentages (nozzle i.d. 7 mm, 200 A, 22 slm Ar *, + 3 , + 5 slm H_2).

Fig. 10 illustrates results obtained for Ar-H_2 plasma jets in this fig., the error bar represents the standard deviation for 50 measurements. The increase of the hydrogen percentage gives more impulse to the jet, with the result that the velocity decreases more slowly. The difference in velocity with pure argon is almost a factor of 3, in good agreement with the ratio of the power levels (also about 3) for the same arc current.

3. IN FLIGHT PARTICULES MEASUREMENTS

Measurement problems are complex :

- the high temperatures (9000 to 15000 K in the plasma jet core) result in very intense plasma emission, which it makes difficult to detect scattered light or light emitted by hot particules /8/. This results in a limitation of size to detect the light scattered by the particules in the plasma core. This size limitation depends on the power of the CW laser that is used to illuminate the particles. For example in typical plasma spray jets (Ar-H_2 25 vol%, 40 kW) a1.5 W argon ion laser line allows one to follow particules down to 1 μm in diameter. For particule surface temperature, the plasma emission makes any measurement in the plasma core impossible.

- the high velocities (up to 400 m/s) and the resulting very short residence times of particules in the small measurement volumes (a few thousands of cubic microns) make it difficult to track the steep gradients /8/.

- the particules have wide trajectory distributions, resulting from their size and injection velocity distributions /48/. Thus, particules having different sizes, velocities and surface temperatures travel in the same measurement volume.

3.1. Qualitative or semi-qualitative measurements

- The oldest techniques are those using a cine-streak photography technique. For example,see the work of Lewis and Gauvin in 1973 /49/ in an atmospheric D.C. jet or, more rencently, the one of Frind et al /50/ to determine statistical values of the velocities of particules in a low-pressure (5.5 10^3 Pa) D.C. plasma jet under supersonic flow conditions.

- Recently, 2D laser imaging has been presented which provides information on the particule distributions within the plasma jet according to the power level, gas flow rate, powder injector location and i.d., carrier gas flowrate, etc... A planar laser sheet (obtained with a cylindrical lens and a frequency doubled Nd-Yag laser) about 300 mm in thickness, is positionned at different distances from the plane defined by the plasma jet and the powder injector /51/. The 90° scattered light is recorded via a computer controlled 2-D array detector and stored in a digital frame storage unit. The background light from the plasma is considerably reduced by using a narrow band pass filter (3 nm) centered at the laser wavelength and appropriate gating (100 ns).

3.2. Quantitative measurements

3.2.1. Velocity, size, surface temperature

In fact the information which is needed is the velocity, surface temperature, and diameter of a single particule at a given position. Such measurements, based on the light pulse method, have been developed by Sakuta and Boulos /52/ in RF plasmas, where particule velocities are below 50 m/s. Their system has never been tested for D.C. plasma jets, but with fast response times (~ 1 ns) it could probably work. A system based on the use of two double wavelength optical fiber pyrometers developed by Vardelle et al /53-56/ allows one to determine the velocity and surface temperature of a single particule in flight in D.C. plasma jets or during its flattening.

Nevertheless all of the particule measurements performed in D.C. plasma spray jets are statistical ones /8, 19,57 to 61/. The velocity is measured by laser velocimetry either using doppler or two focus methods. The flux number density is determined by counting the number of signals obtained from the particules crossing the velocity measurement volume. The particule diameter measurement is based on Mie theory, correlating for example the laser doppler signal pedestal amplitude to the diameter of the particule, which is assumed to be spherical /62/. However, the trajectory ambiguity (particules passing through the center or the edges of the measurement volume of the laser doppler anemometry) has to be removed /61/. Because of this ambiguity, simultaneous measurements of the size and velocity distributions are achieved, but not simultaneous measurements of the size and velocity of single particules. Of course, in all the cases the measurement volume can be moved across the plasma jet to obtain radial distributions either in one direction or in two orthogonal ones. For example, in the last case this allows one to determine a 2D particules distribution in a cross section of the plasma jet at a given stand-off distance. The thermal history of the particules is monitored by two color pyrometry. However, the interpretation of the results must be made cautiously :

- the measurement volumes for velocity, number density, and size on the one hand as compared to surface temperature on the other hand are not necessarily the same.

- the counting can also be quite different for the various measurements. If pratically all the particules passing through the measurement volume are taken into account for flux measurements, some signals for laser velocimetry or size can be rejected and the corresponding mean values are calculated for a different number of particles than that used for the flux. For temperatures below 2000 K no signal is detected and thus only the hot particules are seen even if the velocity and/or size of the cold ones is measured. It is, however, possible to take into account only the size of the particules giving a temperature signal /61/. In spite of all these drawbacks, such measurements are very useful. For example fig.11, illustrates the influence of the arc current on the axial velocity and temperature profiles for carbide particules in an Ar-H_2 plasma 100 mm downstream of the nozzle exit along the mean trajectory. It can be seen that the velocity increases by 40 m/s (35%) when the arc current is raised from 350 to 650 A, and the temperature increase by 400 K i.e. only 10% (in good agreement with the decrease of the residence time).

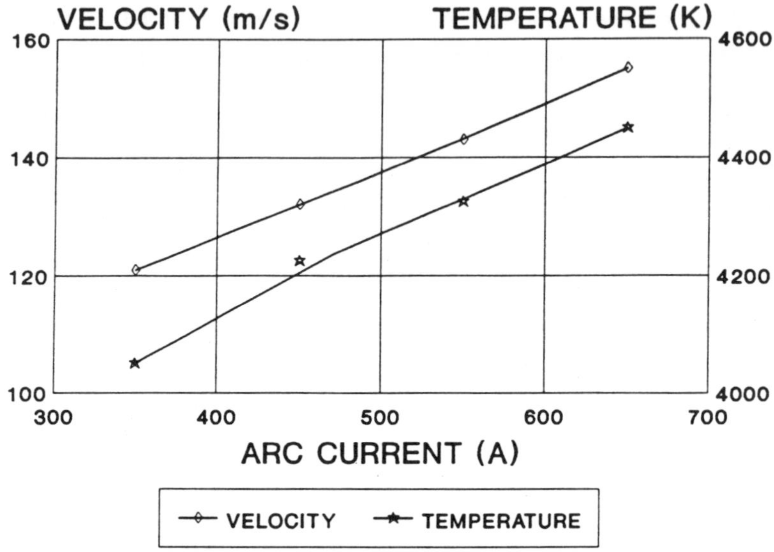

Fig. 11 : Evolution with the arc current of the velocity and surface temperature of carbide particules injected in an Ar-H_2 plasma jet (72 slm Ar, 18 slm H_2). The measurements were taken 100 mm downstream of the nozzle exit along the mean particule trajectory

Such a low increase in temperature when the gas enthalpy is raised by almost 90% is due to the fact that the length of the plasma jet increases very little (see fig. 9) while its velocity is almost doubled, thus reducing the residence time of the particules. Notice however that in the special case of these heavy carbide particules, the increase in their velocity is only 27% against 100% for the plasma.

In fig.12, the constriction of the plasma jet when the gas flowrate is increased can be seen. In spite of a higher velocity with a higher flowrate, the surface temperature of the particules is increased. The result is very clear here because a long nozzle with gradually increasing diameter was used to reduce pumping of the surrounding air. Without this extended nozzle and with an external powder injection, the increased gas flowrrate induces more air pumping, thus cooling down the plasma jet even more rapidly and surface temperatures are almost the same for both flow rates.

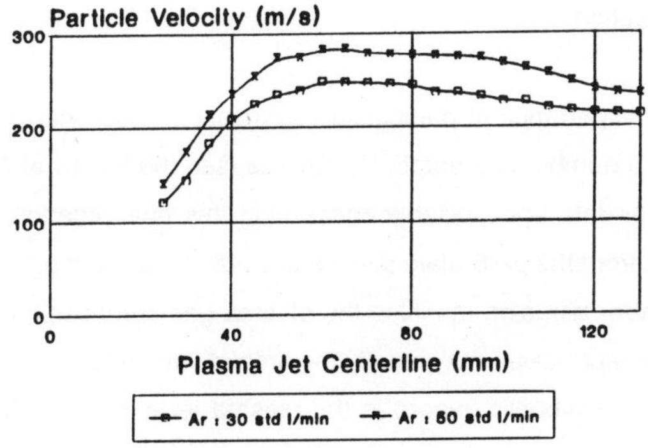

a)

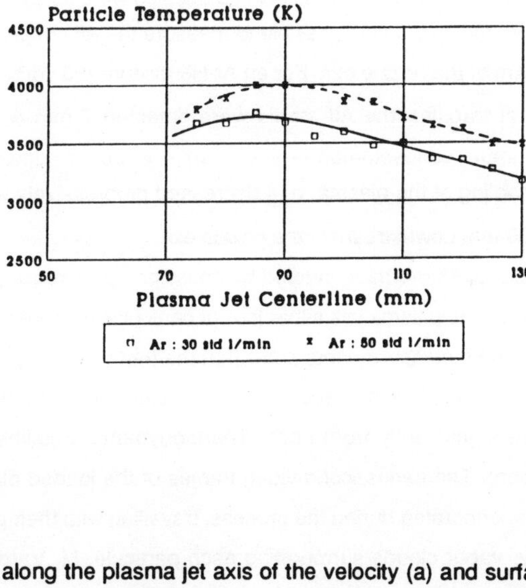

b)

Fig. 12 : Evolution along the plasma jet axis of the velocity (a) and surface temperature (b) of yttria particules injected in the diverging region of a long PTF4 plasma torch (nozzle i.d. 7 mm, 440 A) for two different plasma gas flowrates a) 30 slm Ar, 12 slm H_2, b) 50 slm Ar, 12 slm H_2.

3.2.3. Vaporization

Although the importance of particulate vaporization under thermal plasma conditions has been stressed in a number of theoretical analyses (see the review of A. Vardelle et al./12/) very few experimental data are available concerning this phenomenon. When vaporization occurs, mass transfer from the particulate surface towards the plasma gas reduces the effective heat transfer. The vapor diffusion modifies the plasma gas composition and so its transport properties. Finally the vapor cloud may modify the droplet flattening.

The presence of metallic vapors in the plasma jet can be evaluated by absorption spectroscopy. This allows one to determine the ground and metastable state populations of metal atoms or ions resulting from the decomposed vapor /63/ (for example Co from Co particules or Al from AlO emitted by Al_2O_3 particules). Measurements are carried out along the mean trajectory of the particules using a hollow cathode lamp as the light source. For example, such measurements have shown that when alumina particules were sprayed at 15 kW with pure argon as the plasma forming gas, it was possible to measure the AlI concentration ($n_{AlI} \sim 10^{16}$ m^{-3}) 20 mm downstream of the nozzle exit. For an Ar-H_2 mixture (20 vol%) with P=17 kW, due to the much better heat transfer, the AlI atoms were detected 7 mm downstream from the nozzle exit with a maximum concentration ($\sim 3.10^{19}$ m^{-3}) at 30 mm downstream. For higher distances, due to the cooling of the plasma, n_{AlI} decreased progressively with densities on the order of 10^{17} m^3 at 100 mm downstream of the nozzle exit.

Detering and Eddy /18,64/ have studied by emission spectroscopy and computerized video image processing Ar-He plasma jets either free of particules or loaded with NiAl or WC-Co powders. The spectroscopic measurements show that the particules (especially W and Co for the cermet particules) are significantly vaporized and ionized, resulting in Ar-He-metal vapor plasmas which deviate significantly from Local Thermodynamic Equilibrium (LTE) with the presence of metal vapors. The nanosecond video frames of the loaded plasma jets show the powder plasma comets, generated during the process, travelling with their particules.

To study these vapor clouds surrounding each particule, M. Vardelle et al /63/ have developed a technique based on emission spectroscopy. The temperature in the "diffusion zone" surrounding one particule is determined by the intensity ratio of the vapor lines simultaneously measured. The vapor radius is deduced from the measurement of the vapor concentration around the particle in flight together the particule velocity (by laser anemometry). The vapor concentration is calculated from the absolute line intensity profile, once temperature is known. Beside the assumptions made under equilibrium conditions in the vapor cloud, such measurements imply symmetrical vapor clouds and can be performed only in regions where the plasma velocity is close to that of particules. The results, at a given location, correspond generally to an average calculated on 20 events. For example fig. 13 shows the Co concentration distributions for WC-Co particules in an Ar-H_2 plasma jet. The radius of the clouds is estimated to be of the order of 1-1.5 mm (one order of magnitude higher than the particule diameter) and corresponds to 10% of the maximum value of the concentration profiles. The temperature is nearly constant within the vapor cloud and is about that of the plasma jet at the same distance from the nozzle exit. The Co concentration is fairly high (~ 10^{23} m^{-3}) near the particule wall. The evolution of the comet radii at different radial locations may be explained by the different trajectories followed by the various particules investigated.

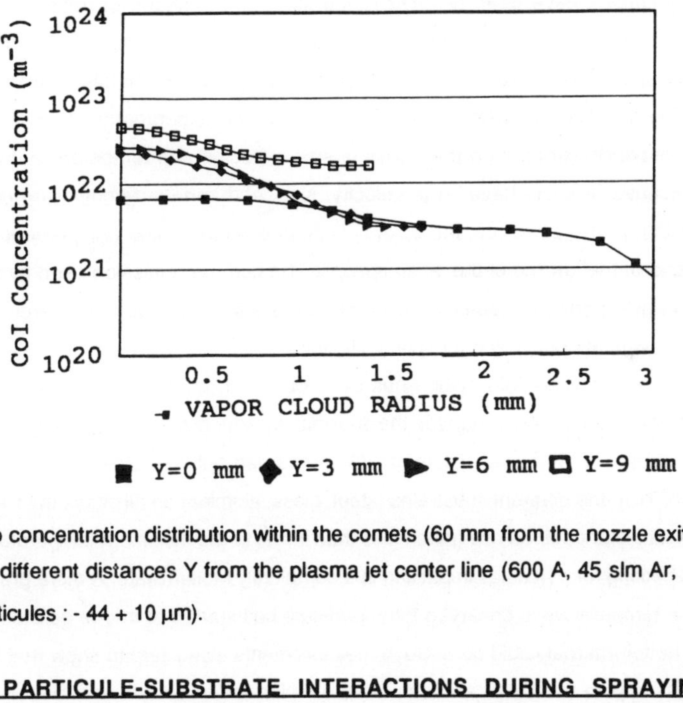

Fig. 13 : Co concentration distribution within the comets (60 mm from the nozzle exit, 7 mm i.d. nozzle) for different distances Y from the plasma jet center line (600 A, 45 slm Ar, 15 slm H_2, WC-Co particules : - 44 + 10 μm).

4. PARTICULE-SUBSTRATE INTERACTIONS DURING SPRAYING

A droplet flattens either on the substrate or on already solidified droplets and experiences radial spreading of the liquid and solidification in times on the order of a few tens of ms /65,66/. According to the powder mass flow rate and the torch to substrate relative velocities, the flattened and solidified droplets form in times of the order of ms a more or less thick sprayed bead, which shape is roughly Gaussian. Then the beads overlap to form passes, the coating being made of a given number of successive passes. The beads and passes are heated by the plasma jet plume and the impacting particules and cooled, if necessary, by air or liquid gas jets. Cooling is of primary importance because the temperature gradients within the coating and the substrate, together with the quenching stresses within the individual lamellae, control the residual stresses and the resulting macro-cracks.

4.1. Study of individual droplets flattening

Many studies have been devoted to the study of the shape of flattened droplets or splats collected on translated or rotated substrates (usually glass slides) crossing the flow /12/. The collected splats exhibit various morphologies from completely circular discs to exploded particules. The simplified models /65 to 67/ as well as the experiments show that the splat morphology depends strongly on the particule size, velocity, and temperature upon impact. However, because droplets have large velocity, size, and surface temperature distributions when impacting on the substrate, the splat collection is tedious, and comprehensive studies are scarce and based on numerous assumptions. The acoustic emission used to monitor the impact of individual particules (see the review of /12/) is also very difficult to interpret according to the same arguments presented for splat collection.

Much more precise information has been obtained due to the development of a two color pyrometric technique to monitor the thermal evolution of individual droplets as they impact on a substrate /54 to 56/. Tests have been carried out with different kinds of sprayed materials (Nb, TiC) and different substrates (steel, glass, alumina). In all cases the thermograms show an almost instantaneous rise (much less than 1 ms) to the maximum temperature followed by a gradual cooling for particules impacting in a completely molten state. A more gradual rise to the maximum temperature is observed if the particule has started to freeze before impact with an unveiled hotter internal liquid core. Such measurements also seem to show that the cooling rate is increased with the substrate temperature which might be due to a better flow of the molten particule on the hotter surface. Experiments related to the whole coating seem to confirm such results. The adhesion and cohesion of alumina coatings on cast iron substrates is increased by a factor of four when spraying on a substrate preheated to 400°C, while maintaining the coating temperature at 400°C during spraying and letting the coating and substrate cooling down slowly /68/. This shows the importance of a better understanding of the parameters controlling the flattening and solidification of droplets.

With a new technique using two pyrometers, developed at this laboratory /12,56/, it is possible to determine the temperature of the particule prior to impact (2 mm ahead of the substrate), its velocity upon impact, its surface and temperature evolutions during impact, as well as its flattening degree. For example, fig. 14 illustrates the time temperature evolution after impact for a 30 μm W particule sprayed onto a steel substrate with an Ar-H_2 plasma jet. The measurement of the droplet surface temperature prior to impact confirms that the particule has started to cool down (at about 60 mm downstream of the nozzle exit as shown by the statistical measurements in flight), and that the core is hotter than the surface. From this thermogram, the cooling rate of this particule during the first ten μs is estimated to be as high as 10^8 K/s.

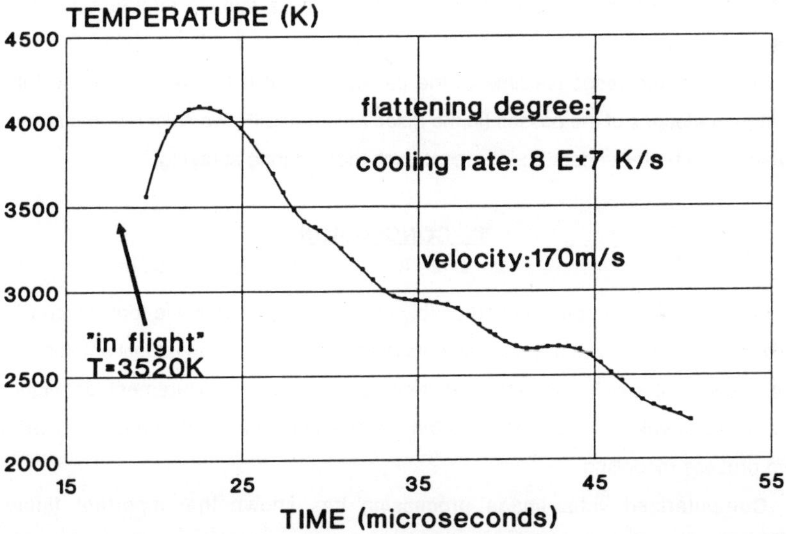

Fig. 14 : Time-temperature evolution after impact for a W —— particule (- 45 + 22 μm) sprayed with an Ar-H_2 plasma jet (45 slm Ar, 15 slm H_2, I=500 A, nozzle i.d. 7 mm, spraying distance 120 mm).

4.2. Study of the coating as a whole during spraying

One of the key parameters to obtain high quality coatings, as outlined in the introduction of this section, is to monitor the temperature of the coating during spraying. This can be achieved through a better knowledge of the heat flux distributions from the particules and the plasma jet. For particules, a few calorimeters have been developed /67/. For the plasma jet itself, assuming the heat flux is Gaussian allows one to determine its distribution (maximum flux and Gaussian radius) /69/.

To monitor the surface temperature, thermocouples located in the substrate can only give values related to the substrate, which may be widely different from the surface temperature (especially when spraying ceramics) /68/. IR thermography permits one, with a previous calibration of the coating emissivity (which may vary in a wide range during spraying due for example to sub-oxides formation), to determine a surface temperature /70/ of the coating.

However, the response time of the camera (about 0.1 s) is too slow to follow the temperature variations of the passes (in the msec range) and IR two color pyrometry has still to be developed to better understand the heat exchanges during spraying.

5. CONCLUSION

Although plasma spraying is probably one of the most versatile tools to apply a wide variety of coatings on almost any substrate, its development has been largely empirical over the last two decades. It is only during the past few years that new measurement techniques have been used to develop better understanding of the complex phenomena involved and to improve process modelling.

Computerized video-image processing has shown the important fluctuations undergone by the plasma jets and underlined the necessity to better understand the coupling between the arc root fluctuations and the cold gas flow in the plasma column fringes. Such measurements, however, imply the development of signal processing techniques to extract characteristic frequencies from noisy voltage, arc current, light intensity and accoustic signals. These fluctuations induce a type of pulsed flow in the plasma jet which contributes to the development of turbulence and surrounding atmosphere pumping. Pumped nitrogen concentration and temperature measurements by CARS, as well as LDA velocity measurements of small particules entrained and injected in the plasma gas, have shown that this pumping is of an engulfment type. Subsequently, the mixing of cold, dense air "bubbles" with the plasma is not as fast as was previously thought.

For the plasma jets, although computerized spectroscopic measurements are now well established, the meaning of the resulting temperatures has to be looked at very carefully due to the fluctuations, the surrounding atmosphere pumping and the non equilibrium effects. These effects are important, even at atmospheric pressure. However, such measurements permit comparison of the influence of the design and working parameters of plasma torches and optimization of them. A comparison of the spectroscopic temperatures with those deduced from enthalpy probes at distances higher than 30 mm from the nozzle exit shows a rather important discrepancy between these measurements. Enthapy probes measure the temperature resulting from the plasma mixing with the surrounding atmosphere. Rayleigh scattering can be used to measure heavy species temperatures in regions where $T < 6000$ K, and this technique could be very useful for measuring heat transfer coefficients to a substrate in connection with heat flux measurements.

Enthalpy probes can also be used to determine the concentrations of plasma gas and surrounding atmosphere in the plasma jets as well as the jet velocities. For velocity measurements, the study of the light fluctuations, with a proper signal treatment, gives access to a simple determination of the axial plasma flow velocity.

For particules in flight, the statistical measurements of velocity, surface temperature, number density, and diameter of particules in flight have confirmed that they follow very different trajectories and undergo wide different heat treatments. However, one has to be aware that such measurements are not necessarily related to the same group of particules even if the same measurement volume and time are considered and their interpretation is not straight forward. 2D laser imaging confirms the trajectory dispersion induced by the size and injection velocity distributions of the particules as well as by the collisions of the particles with the injector wall resulting in non negligible radial velocity distributions at the injector exit. This technique can be used to great advantage to optimize the powder injector location and injection angle. Particule vaporization, especially for particules in which the heat propagation is not negligible, is important. The spectroscopic measurements have shown, in regions where vapor clouds are symmetrical (where particle velocity is close to that of the plasma), the importance of their size (an order of magnitude greater than that of the particule). Such big vapor clouds modify the heat transfer between the plasma and the particle. However techniques are still be developed to study these vapor clouds close to the substrate and to evaluate their importance for particule flattening. It should also be noticed that the plasma jet length fluctuations, revealed by computerized video image processing raise the question of particules treatment : should it be continuous, as commonly allowed, or are the particules heated more or less according to their time of passage in the fluctuating plasma jet ?

Finally the splats collected at different distances and radii downstream of the nozzle exit, exhibit a wide variety of shapes and morphologies. Recently developed pyrometric techniques permit the measurement of particle velocity and temperature prior to impact, as well as the evolution of their temperature and surface (flattening degree) during flattening. They will probably contribute to a better understanding of how the splats are formed, and the resulting contacts between the new incoming splat and the previously deposited ones or the substrate. Such systematic measurements are of primary importance to control the thermomechanical properties of the coatings. However such control also implies the development of new fast (~ 10 kHz) IR pyrometers to monitor the surface temperature of beads and passes during spraying.

REFERENCES

/1/ N.N. RYKALIN, V.V. KUDINOV, Pure and Appl. Chem. 48, p. 299 (1976).

/2/ J.H. ZAAT, Ann. Rev. Mater. Sci. 13, p. 9 (1983).

/3/ Thermal spraying, Practice, Theory and Application (1985) (Pub.) Am. Welding Soc. P.O. Box 35, 1040 Miami, Fl, USA.

/4/ Technology Forecast, Advanced Materials and Processes 133, 1, p. 8 (1988).

/5/ P. FAUCHAIS, A. GRIMAUD, A. VARDELLE, M. VARDELLE, Ann. Phys. Fr. 14, p. 261 (1989).

/6/ J.F. FINKE, R. RODRIGUEZ, C.G. PENTECOST, p. 45 in Thermal Spray Research and Applications (1980) (Pub.) ASM Int. Materials Park, Ohio, USA.

/7/ R. SPORES, E. PFENDER, Surface and Coating Technology 37, p. 251 (1989).

/8/ P. FAUCHAIS, J.F. COUDERT, M. VARDELLE, p. 349 in Plasma Diagnostics (1989) (Pub.) Academic Press Inc.

/9/ J. DONG, R.J. KEARNEY, Plasma Gas Velocities in an Argon Jet at Atmospheric Pressure, submitted to plasma chemistry and plasma processing.

/10/ J.R. FINKE, W.D. SWANK, p. 39 in Thermal Spray Research and Applications (1990) (Pub.) ASM Int. Materials Park, Ohio, USA.

/11/ M.I. BOULOS, Proc. of 2nd Plasma Technik Symposium 1 p. 49, (1991).

/12/ A.VARDELLE, M. VARDELLE, P. FAUCHAIS, Diagnostics for particulate vaporization and interactions with surfaces, accepted in Pure and Applied Chemistry.

/13/ P. FAUCHAIS, J.F. COUDERT, A. VARDELLE, M. VARDELLE, A. GRIMAUD, P. ROUMILHAC, p. 11 in Advances in Coating Technology (1988), (Pub.) ASM Int., Metals Park, Ohio, USA.

/14/ J.F. COUDERT, C. DELALONDRE, P. ROUMILHAC, O. SIMONIN, P. FAUCHAIS, Modelling and experimental study of a transferred arc stabilized with argon in an argon atmosphere, submitted to Plasma Chemistry Plasma Processing.

/15/ P. ROUMILHAC, Ph.D. Thesis, University of Limoges March (1990).

/16/ J.M. LEGER, Ph.D. Thesis, University of Limoges March (1990).

/17/ E. PFENDER, W.L.T. CHEN, R. SPORES, p. 1 in Thermal Spray Research and Applications (1990) (Pub.) ASM Int., Materials Park, Ohio, USA.

/18/ B.A. DETERING, J.R. KNIBLOC, T.L. EDDY, p. 27 in Thermal Spray Research and Applications (1990) (Pub.) ASM Int., Materials Park, Ohio, USA.

/19/ J.R. FINCKE, C.L. JEFFERY, S.B. ENGLERT, Mat. Res. Soc. Symp. Proc. 98 p. 127 (1987).

/20/ T. HOFFMAN, Advanced Materials and Processes 140 (3) p. 37 (1991).

/21/ T.L. EDDY, J.D. GRANDY, B.A. DETERING, Heat Transfer in Thermal Plasma Processing HTD 161 p. 57 (1991) ASME, NY, USA.

/22/ J.F. COUDERT, P. FAUCHAIS, Plasma flow velocity measurements in d.c. arc jets using cross correlation analysis of the luminuous fluctuations to be submitted to Plasma Chemistry, Plasma Processing.

/23/ J.R. FINCKE, R. RODRIGUEZ, C.G. PENTECOST, Mat. Res. Soc. Symp. Proc. 190, p. 207 (1991).

/24/ E. PFENDER, J. FINCKE, R. SPORES, Plasma Chemistry Plasma Processing 1 p. 529 (1991).

/25/ R. SPORES, E. PFENDER, Surface and Coating Technology 37 p. 251 (1989).

/26/ R. SPORES, Ph.D. Thesis, Dpt. of Mechanical Eng., Univ. of Minnesota Minneapolis, MN, Dec. (1989).

/27/ S.C. SYNDER, J.D. GRANDY, HTD, Heat Transfer in Thermal Plasma Processing 161, p. 95 (1991) (Pub.) ASME, N.Y., USA.

/28/ J.F. COUDERT, P. FAUCHAIS, The influence of the arc fluctuations on the temperature measurements in D.C. plasma jets (1992) TSM annual meeting.

/29/ T.L. EDDY, J.M. LEGER, J.F. COUDERT, P. FAUCHAIS, Heat Transfer in Thermal Plasma Processing, HTD 161 p. 67 (1991) ASME, NY, USA.

/30/ T.L. EDDY, B.A. DETERING, J.A. BATDORF, P.G. SCHENK, Heat Transfer in Thermal Plasma Processing HTD 161 p. 57 (1991) ASME, NY, USA.

/31/ P. ROUMILHAC, J.F. COUDERT, P. FAUCHAIS, Mat. Res. Soc. Symp. Proc. 190 p. 227, (1991).

/32/ A.C. ECKBRETH and T.J. ANDERSON, Applied Optics 24 P. 2731 (1985).

/33/ R.R. AUTCLIFF, T.J. ANDERSON, Applied Optics 24, p. 2731 (1985).

/34/ R.J. HALL, A.C. ECKBRETH, CARS : application to combustion diagnostics, in Laser Applications, Vol. 5, (Eds.) J.F. READY, R.K. EAF, (Pub.) Academic Press, NY (1984).

/35/ A.J.D. FARMER, G.H. HADDAD, Journal of Physics D : Applied Physics 22, p. 426 (1988).

/36/ Z. WANG, R.J. KEARNEY, J.Q.S.R.T. 44, p. 39 (1990).

/37/ J.R. FINCKE, C.G. PENTECOST, HTD, Heat Transfer in Thermal Plasma Processing 161, p. 101, (1991) (Pub.) ASME, NY, USA.

/38/ M. BROSSA, E. PFENDER, Plasma Chemistry, Plasma Processing 8 (1) p. 75 (1988).

/39/ Y.P. CHYOU, E. PFENDER, Plasma Chemistry, Plasma Processing 9 (2) p. 291 (1989).

/40/ M. VARDELLE, A. VARDELLE, P. FAUCHAIS, M. BOULOS, AIChE 29, P. 236 (1981).

/41/ J. LESINSKI, M. MIZERA-LESINSKA, J.C. FANTON, M. BOULOS, AIChE 27, p. 252 (1981).

/42/ E. PFENDER, Pure and Applied Chemistry 57 (9) p. 1179 (1985).

/43/ R. SPORES, E. PFENDER, p. 85 in Thermal Spray Technology, New Ideas and Processes (Pub.) ASM Int., Oh, USA (1989).

/44/ W.L. BOHM, M.U. BETH, G. NEDDA, J.Q.S.R.T. 7 p. 661 (1967).

/45/ A.W. KOCH, K.D. LANDES, G. SEEGER, W. TILLER, Local measurements of gas velocity in an argon plasma jet, Int. Institute of Welding, Germany (1989).

/46/ A. CAPETTI, E. PFENDER, Plasma Chem. Plasma Proc. 9 (2) p. 329 (1989).

/47/ J.D. GRANDY, R.J. KEARNEY, HTD, Heat Transfer in Thermal Plasma Processing 161, p. 107 (1991).

/48/ P. FAUCHAIS, M. VARDELLE, A. VARDELLE, J.F. COUDERT, Metallurgical Transactions 20B p. 263 (1989).

/49/ J.A. LEWIS, W.H. GAUVIN, AIChE 19 p. 982 (1973).

/50/ G. FRIND, C.P. GOODY, L.E. PRESCOTT, ISPC6, 1 p. 120 (1983) M. BOULOS, Univ. of Sherbrooke, CN.

/51/ W.C. ROMAN, M. WINTER, A.A. ROTUNNO, K. FESSENDEN, W. WILLER, Thermal Spray Research and Applications, p. 49 (1991) (Pub.) ASM Int., Oh, USA.

/52/ T. SAKUTA, M. BOULOS, ISPC9, 1 (1989) p. 371.

/53/ M. VARDELLE, State thesis, University of Limoges, July 87.

/54/ C. MOREAU, P. CIELO, M. LAMONTAGNE, S. DALLAIRE, M. VARDELLE, Meas. Sci. Technol. 1 p. 807 (1990).

/55/ C. MOREAU, P. CIELO, M. LAMONTAGNE, S. DALLAIRE, J.C. KRAPEZ, M. VARDELLE, p. 19 in Thermal Spray Research and Application (1991) (Pub.) ASM, Oh, USA.

/56/ C. MOREAU, M. LAMONTAGNE, P. CIELO, Influence of the coating thickness on the cooling rates of plasma sprayed particles impinging on a substrate, NTSC 91. To be published by ASM.

/57/ A. VARDELLE, J.M. BARONNET, M. VARDELLE, P. FAUCHAIS, IEEE PS8 p. 417 (1980).

/58/ A. VARDELLE, M. VARDELLE, P. FAUCHAIS, Plasma Chemistry Plasma Processing 2 p. 255 (1982).

/59/ J. MISHIN, M. VARDELLE, J. LESINSKI, P. FAUCHAIS, J. of Phys. E. Sci. Instr. 20 p. 620 (1987).

/60/ M. VARDELLE, A. VARDELLE, P. FAUCHAIS, M. BOULOS, AIChE 34 (4) p. 567 (1988).

/61/ J.R. FINCKE, C.L. JEFFERY, National Thermal Spray Conf. Proc. p. 55 (1988) (Pub.) ASM, Oh, USA.

/62/ G. GOUESBET, Plasma Chem. Plasma Proc. 5 (2) p. 91 (1985).

/63/ M. VARDELLE, A. VARDELLE, C. TRASSY, P. FAUCHAIS, Plasma Chem. Plasma Proc. 11 (2) p. 185 (1991).

/64/ T.L. EDDY, B.A. DETERING, G.C. WILSON, p. 33 in Thermal Spray Research and Applications (1990) (Pub.) ASM Int., Metals Park, Oh, USA.

/65/ J. MADJESKI, Int. J. Heat Mass Transfer 19 p. 1009 (1766).

/66/ R.Mc. PHERSON, J. of Materials Science 15 p. 3141 (1980).

/67/ J.M. HOUBEN, Ph.D. Thesis, University of Eindhoven, N.L. (1988).

/68/ L. GYENIS, A. GRIMAUD, O. BETOULE, F. MONERIE-MOULIN, P. FAUCHAIS, M. DUCOS, 2nd Plasma-Technik Symposium 1 p. 95 (1991) (Pub.) P.T. Wohlen, CH.

/69/ F. MONERIE-MOULIN, F. GITZHOFFER, P. FAUCHAIS, M. BOULOS, Study of the heat flux transmitted by a D.C. Ar-H_2 plasma spraying jet to a cold substrate, TSM Annual meeting, March 1992, San Diego.

/70/ L. PAWLOWSKI, M. VARDELLE, P. FAUCHAIS, Thin Solid Films 94 p. 307 (1982).

Chapter 3

EXPERIMENTAL DESIGN AND OPTIMIZATION OF PLASMA SPRAYED COATINGS

Shao L. Chen, Pekka Siitonen, Pentti Kettunen
Institute of Materials Science
Tampere University of Technology
P.O. Box 589, SF-33101 Tampere, Finland

Abstract

Orthogonal block design, regression analysis, and optimization of spraying variables (ORO) is applied to achieve a quantitative and optimized prediction of the structure or properties of plasma sprayed coatings. The orthogonal block design is used to arrange the experiments and to reduce their number. The received structure or property is a function of the selected variables, but additionally also of the interactions between them. Applying regression analysis, direct relation between spraying variables and the coating structure or property are obtained as second order equations. The successive quadratic programming algorithm is then used on the obtained equations to find the variables corresponding to the most suitable property. The validity of the equations is experimentally checked at the point of optimum and outside it. Spraying with the optimized variables leads to a significant improvement in coating properties.

1. Introduction

Plasma-sprayed coatings are widely used in corrosion and wear protection, or thermal and electrical insulation, etc. A variation in spraying variables affects, however, the structure and properties of the received coatings, even in the case of identical materials and equipment [1-3]. Quality assurance, i.e., a repeatable achievement of high quality coatings, implies the finding of correct values of multiple variables for many different cases. Successful results are reached, e.g., in prediction of particle temperature and velocity during the flight in plasma stream by mathematical modeling based on the principles of heat transfer and two-phase flow [4,5]. Still the method is insufficient for quantitative prediction and direct optimization of coating properties.

Similarly, orthogonal design at two levels is used to arrange the experiment in thermal spraying, by which the significance of spraying variables is evaluated [6-8]. In spite of these positive results, a practical method for quantitative prediction and optimization of coating properties (e.g., porosity, hardness, wear rate, etc.) or spraying parameters (e.g., power, voltage, powder and power efficiencies, etc.) is not developed until yet.

The present paper introduces the method of ORO, in which the principles of orthogonal block design, regression analysis, and optimization are in combined form applied to quantitative prediction and optimization of different plasma sprayed coatings.

2. Principles of ORO

The application of ORO proceeds in three steps. First, the orthogonal design is applied to arrange experiments so that the necessary number of them is drastically reduced [9,10], bearing in mind that the properties to be evaluated are the response functions of the selected variables, but also of the interactions appearing between them. This, on its side, depends on the design table selected. Secondly, the discrete experimental data are converted to quantitative equations by means of regression analysis [11]. Finally, on the bases of the obtained equations, the response function of the selected property is optimized by gradient algorithm, successive quadratic programming algorithm or any other suitable method, and the validity of the equations are experimentally checked at the optimal point and outside it.

2.1. Orthogonal Design

In principle, a relationship between the variables, i.e., $\mathbf{z} = (z_1, z_2, \ldots, z_p)^T$ and their response function, $y(z_1, z_2, \ldots, z_p)$, can be expressed by a Taylor polynomial, that is,

$$y = c_o + \sum_{j=1}^{p} c_j z_j + \sum_{i<j}^{p} c_{ij} z_i z_j + \sum_{j=1}^{p} c_{jj} z_j^2 + \ldots \qquad (1)$$

In plasma spraying, the response function can be classified as:

1) the coating properties, such as wear rate, hardness, adhesion strength, etc.;
2) the coating structure, such as porosity, phase composition; and
3) the powder yield and spraying productivity, i.e., spraying time/coating volume.

In plasma spraying, the relations between the spraying variables and the response function appear normally as non-linear and even as non-monotone ones. Therefore, the notification of the interactions and the higher order terms is important and supporting the selection of regression equation of higher order. Consequently, an equation of higher order gives better resolution, but results in a rapid increase of the number of experiments. As a compromise, an equation of second order is normally applied. In order to obtain the second order regression equations in the subsequent step, at least three factor levels are required. This, however, again increases the number of necessary experiments. Basicly, the number of experiments (*n*) should be bigger than the number of coefficients in the equation (*q*). In the present work, an orthogonal block design table is applied which contains three major levels and two extended levels [9], but still requires less experiments than a normal three level orthogonal design. This orthogonal block design table is given in the Appendium 1.

The variation range for each variable is defined by

$$z_{il} < z_i < z_{iu} \qquad\qquad i = 1,2, \ldots, p$$

where, p is the number of spraying variables, z_{il} and z_{iu} are values of ith spraying variable at low and upper levels, respectively. The specification of a plasma equipment usually sets the limits for each variable. Enough leeways to the limitations are always necessary to avoid spray operation beyond them. Without any knowledge, e.g., spraying a new material or using a new equipment, the variation range can be selected by referring to the specification. At the beginning, a wide variation range is usually used in order to reach a high probability for finding the optimal parameters inside it or near it. But such a range selection sacrifices the resolution of the regression equation. To achieve high resolution, a new and narrow range may be selected around the optimal parameters obtained in first ORO run. Since the number of the variables and/or their levels are generally reduced after first ORO run, the number of experiments in the second ORO run is thus reduced. In most practical cases of first ORO run, the variation range of variables can be roughly estimated as based on earlier results of the same or similar material using the same or similar equipment.

In orthogonal design table given in the Appendium 1, the factors are at the levels of +1, 0 and -1, etc. Spraying variables ($z_1, z_2, \ldots, z_p$) need to be indexed into factors ($x_1, x_2, \ldots, x_p$) corresponding to the design table. For variable z_i varying in a range from z_{il} to z_{iu}, the corresponding factor x_i varies in the range of -1 and +1. If z_i at 0 level is given by

$$z_{i0} = (z_{il} + z_{iu}) / 2 \ , \qquad (2)$$

then x_i related to z_i yields

$$x_i = (z_i - z_{i0})/(z_{iu} - z_{i0}) \qquad (i = 1,2,\ldots,p) \qquad (3)$$

Eq. 1 then assumes the form

$$y = b_0 + \sum_{j=1}^{p} b_j x_j + \sum_{i<j}^{p} b_{ij} x_i x_j + \sum_{j=1}^{p} b_{jj} x_j^2 \tag{4}$$

where, b_0 is constant of the equation, b_i, b_{ij} and b_{jj} are coefficients of the equation. Note that the second order terms ($x_1^2, x_2^2, \ldots, x_p^2$) in the orthogonal design table are replaced by $x'_1, x'_2, \ldots, x'_p$ for the reason of orthogonality. Furthermore, the elements x'_{kj} in the column of x'_j are related to element x_{kj} of x_j^2 according to the rule:

$$x'_{kj} = x_{kj}^2 - \frac{1}{n} \sum_{k=1}^{n} x_{kj}^2 \tag{5}$$

with $i = 1, 2, \ldots, p$
and $k = 1, 2, \ldots, n$

2.2. Regression Analysis

The elements in the orthogonal block design table form a matrix **X** with the size of $n \times (q+1)$, as q is the number of the coefficients. If response function vector **y** and coefficient vector **b** are defined as $\mathbf{y} = (y_1, y_2, \ldots, y_n)^T$ and $\mathbf{b} = (b'_0, b_1, b_2, \ldots, b_p, b_{12}, b_{13}, \ldots, b_{p\text{-}1p}, b_{11}, b_{22}, \ldots, b_{pp})^T$, where, $y_1, y_2, \ldots, y_n$ are measured or recorded data corresponding to each run, then **b** can be obtained by

$$\mathbf{b} = (\mathbf{X'X})^{-1}\mathbf{X'y} \tag{6}$$

and b_0 in Eq. 4 is given by

$$b_0 = \frac{1}{n} \left(\sum_{k=1}^{n} y_k - \sum_{k=1}^{n} x_{kj}^2 \sum_{j=1}^{p} b_{jj} \right) \tag{7}$$

The significances of the obtained equation and each coefficient have to be evaluated by *F*-test [16]. *F*-value for the equation is given by

$$F_{eq} = (SS_{reg} / f_r) / (SS_{res} / f_s) \tag{8}$$

where, SS_{reg} and SS_{res} are the sums of regression error square and residual error square, respectively; and correspondingly f_r and f_s are the regression freedom and the residual freedom, respectively, given by

$$f_r = q \tag{9}$$

$$f_s = n - q - 1 \tag{10}$$

SS_{reg} is calculated by

$$SS_{reg} = \sum_{j}^{q} Q_j \qquad (11)$$

where,

$$Q_j = \frac{\sum_{k}^{n} x_{kj} y_k}{\sum_{k}^{n} x_{kj}^2} \cdot \sum_{k}^{n} x_{kj} y_k \qquad (12)$$

Correspondingly,

$$SS_{res} = SS_{tot} - SS_{reg} \ , \qquad (13)$$

as

$$S_{tot} = \sum_{k}^{n} y_k^2 - \frac{1}{n} (\sum_{k}^{n} y_k)^2 \qquad (14)$$

If F_{Eq} is bigger than F_α (f_r, f_s), it means that the equation is with confidence at a level of (1-α). The F_j for each coefficient is given by

$$F_j = Q_j / (SS_{res} / f_s) \qquad j = 1, 2, \ldots, q \qquad (15)$$

If the F_j-value of a coefficient is the lowest one among the F_j-values of other coefficients and less than standard $F_{\alpha=0.25}$ at the same degrees of freedom, the corresponding term is deleted from the equation. Then, the F_j-values have to be calculated again and this checking procedure is repeated until all the values of remaining terms are greater than the standard $F_{\alpha=0.25}$.

2.3. *Optimization*

The optimization problem is to find the minimum of the response function, that is

$$\min y = f(x_1^*, x_2^*, \ldots, x_p^*)$$

with constraints

$$y_c(x_1, x_2, \ldots, x_p) > y_{co} \qquad c = 2, 3, \ldots, m$$

$$a_i < x_i < b_i \qquad i = 1, 2, \ldots, p$$

where, $x_1^*, x_2^*, \ldots, x_p^*$ donate the factorial values which correspond to the optimal spraying variables; $y_c(x_1, x_2, \ldots, x_p)$ = constraining functions, y_{co} = limitation values, m = the number of constraints, a_i and b_i are low and upper limitations of $x_1, x_2, \ldots, x_p$, respectively. $y_c(x_1, x_2, \ldots, x_p)$ can be power, and/or coating properties, providing that the corresponding regression equations are obtained. The optimization problem can be solved by using a gradient algorithm [12], or a successive quadratic programming algorithm [13], as done, e.g., in the present work. The problem for searching the maximum of the response function can be solved by minimizing (-*y*).

3. Application to alumina coatings

The method of ORO was first applied to spraying of alumina coatings, especially to minimize the porosity of the sprayed layer, but also to optimize the wear resistance and the powder yield.

3.1. Experimental details

Alumina powder used was commercial Amperit 740.1 (fused and crushed powder), with a particle size range of 5.6 μm - 45 μm. X-ray diffraction analysis proved that the powder was mostly constituted of α-alumina.

The coatings were prepared by a plasma spraying unit of the type PLASMA-TECHNIK A-3000S using argon and hydrogen as plasma gases. The variables to be optimized were distance from torch to substrate (z_1), current (z_2), argon flow rate (z_3), hydrogen flow rate (z_4), and powder feed setting (z_5) which is proportionally related to the powder feed rate. These variables are indexed as the factors with five levels as listed in Table 1. The button-like substrates were fixed on the lateral surface of a cylinder (200 mm in diameter) which was rotating during spraying at a rate of 120 rpm. The other spraying variables were fixed and are listed in Table 2. Coating thicknesses were controlled in the range of 350 - 450 μm.

Table 1. Index of the spraying variables.

x_i	-1.547	-1	0	1	1.547
Spray distance z_1, [mm]	89	100	120	140	151
Current z_2, [A]	410	450	525	600	641
Ar flow rate z_3, [l/min]	34.2	38	45	52	55.8
H_2 flow rate z_4, [l/min]	3.8	6	10	14	16.2
Powder feed setting z_5, [%]	24.2	28	35	42	45.8

Table 2. Plasma spraying constants.

Plasma torch	F4
Nozzle/anode diameter	6 mm
Powder injector diameter	1.5 mm
Inject distance	6 mm
Carrier gas flow rate	4.2 l/min
Gun speed	2.7 mm/sec

X-ray diffractometer with Cu Kα radiation and a secondary crystal monochromator was employed to identify the phase structures of the powder and sprayed coatings. Optical micrography was used to examine the cross sections of sprayed coatings. The coating porosity was measured using water impregnation method, and rechecked using potentiostatic electrolytic method and image analysis [14]. The abrasive wear resistance of sprayed coatings were determined using the dry sand/rubber wheel method [15]. The surface speed of the wheel was 1.64 m/sec and the specimens with a diameter of 25 mm were pushed against the wheel with a force of 1 kp. A sand curtain at a flow rate of 28 g/min was formed between the specimen and the wheel.

3.2. *Results and Discussion*

3.2.1. *Regression equations and F-test*

The values of recorded power, measured porosity, wear weight loss and calculated powder yield, which are listed in Appendium 2 (Table 1), were used for further regression analysis applying Eqs. 6 and 7. The coefficients in the equations of porosity (y_1), wear weight loss rate (y_2), power (y_3), and powder yield (y_4) were obtained, respectively.

F-values of each equation and of each coefficient were calculated, and compared to *F*-distribution with (f_r, f_s) degree of freedom [16]. The significance of individual coefficient in the above equations is decided by its *F*-value. If F is less than $F(1, f_s)_{0.25}$, the term is deleted. On this basis the final equations of y_i (i = 1, 2, ..., 4) are:

$$y_1 = 8.81 + 0.37x_1 - 0.67x_2 + 0.40x_3 - 0.63x_4 + 0.24x_1x_2 - 0.20x_1x_4 - 0.26x_2x_5 + 0.1x_3x_4 + 0.15x_3x_5 + 0.1x_4x_5 + 0.23x_1^2 - 0.23x_2^2 + 0.27x_4^2 + 0.25x_5^2 \quad (16)$$

with $F = 26.0 > F(14, 12)_{\alpha=0.01} = 4.05$,

$$y_2 = 7.61 + 0.85x_1 - 1.83x_2 + 0.94x_3 - 2.33x_4 + 0.41x_1x_2 + 0.39x_1x_3 - 0.75x_1x_4 - 0.68x_2x_3 + 0.43x_2x_4 - 0.86x_3x_4 + 0.50x_4x_5 + 0.47x_2^2 + 1.57x_4^2 + 0.36x_5^2 \quad (17)$$

with $F = 32.0 > F(14, 12)_{\alpha=0.01} = 4.05$,

$$y_3 = 41.2 - 1.73x_1 + 3.86x_2 - 11.88x_3 + 8.84x_4 + 0.70x_5 - 1.56x_1x_2 - 0.88x_1x_3 + 1.56x_1x_4 - 0.71x_2x_3 + 0.68x_2x_5 + 1.18x_4x_5 + 1.37x_3^2 - 2.45x_4^2 - 2.47x_5^2 \quad (18)$$

$F = 103.5 > F(14, 12)_{\alpha=0.01} = 4.05$.

$$y_4 = 32.7 + 0.33x_1 + 4.78x_2 + 0.57x_3 + 3.46x_4 + 0.3x_1x_3 + 0.25x_2x_4 + 0.62x_1^2 + 0.6x_2^2 + 0.47x_3^2 \quad (19)$$

with $F = 289 > F(9, 17)_{\alpha=0.01} = 3.77$.

The results of *F*-test mean that the obtained equations were all with confidence above the level of (1 - 0.01). The porosity, weight loss rate in wear, powder yield, and power calculated by inputting the factors used in the spraying revealed rather good agreement with experimental results. However, the application of these equations in regions outside the tested area is less reliable.

3.2.2. *Example: Effect of spraying variables on coating porosity*

Under the condition that other variables are at about optima, flow rates of H_2 and Ar are the two determining variables for coating porosity. The optimum appears to be at reasonably small Ar flow rate and rather high H_2 flow rate, as is illustrated in Fig. 1.

Under similar conditions, the optimal porosity appears at reasonably small spraying distance and rather high powder feed, as shown in Fig. 2. The effect of powder feed comes mainly through the interaction terms and the second order term x_5^2, as the *F*-test gave insignificant value for the first order term x_5.

As a function of simple variable, e.g., current, the porosity appears to be a more straight forward function, as shown in Fig. 3.

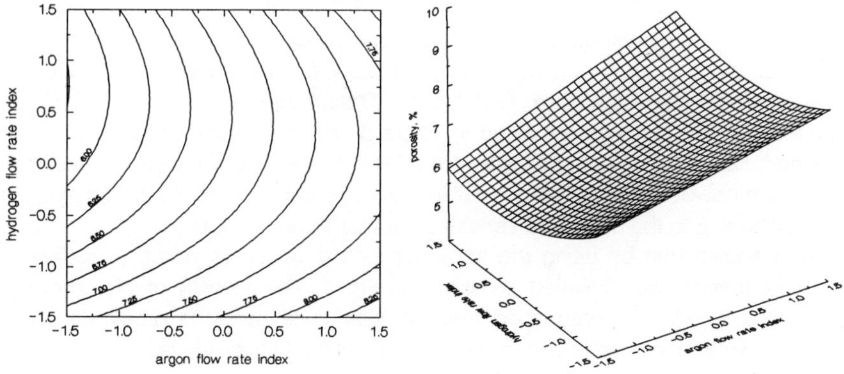

(a) (b)

Figure 1. Porosity as a function of flow rate indices of argon and hydrogen under the condition that the other factors are at optimal levels,
a) contours in two dimensional expression, and
b) three dimensional illumination.

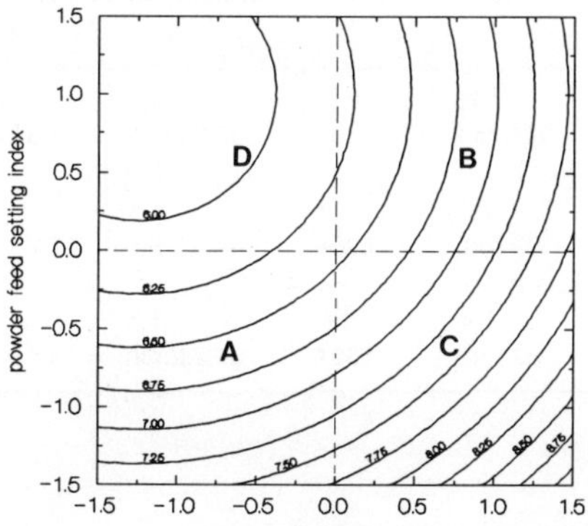

Figure 2. Porosity variation as a function of indices of spraying distance and powder feed setting under the condition that the other factors are at optimal levels.

3.2.3. Optimization of the variables and sprayed coatings

As for individual response function, the optimization was carried out under the condition that the upper and lower boundaries of each factor were determined as a_i = -1.4 and b_i = 1.4 (i = 1, 2, ..., 5). Optimization predicts the spraying variables in the test area as listed in Table 3 and the values of the response function, i.e., the minimum porosity, the minimum wear weight loss rate and the maximum powder yield as listed in Table 4. As comparison, the measured results corresponding to the optimized variables are also listed in Table 4.

It is shown that by using the variables for the minimum porosity, a dense alumina coating was achieved, as shown in Fig. 5. The porosity of this coating, measured by water-impregnation method, was quite near the predicted one. Also, the wear minimized coating gives a wear weight loss rate value almost the same as the predicted. Other recorded or measured values of the response function (power and powder yield) also agree with those predicted. X-ray diffraction analysis showed that both coatings were predominantly constituted of γ phase with a small amount of α-phase.

Table 3. Optimized variable indices x_i and z_i, 1 = spraying distance, 2 = current, 3 = argon flow rate, 4 = hydrogen flow rate, and 5 = powder feed setting

	i	1	2	3	4	5
for minimum	x_i	-1.26	1.4	-1.4	0.71	1.18
porosity,	z_i	94.8 mm	630A	35.2 l/min	12.8 l/min	43.3 %
for minimum	x_i	-1.4	1.4	1.4	0.68	-0.47
wear rate,	z_i	92 mm	630A	54.8 l/min	12.7 l/min	31.7 %
for maximum	x_i	-1.4	1.4	-1.4	1.4	0.66
powder yield,	z_i	92 mm	630A	35.2 l/min	15.6 l/min	39.6 %

Table 4. Comparison of predicted and obtained results.

Optimized	power, kW		porosity, %		wear rate, mg/min		powder yield, %	
for	predicted	recorded	predicted	measured	predicted	measured	predicted	obtained
porosity	44.3	42.8	5.83	6.72	6.88	3.17	73.7	63.7
wear rate	44.8	41.8	7.50	8.14	2.55	2.66	40.2	39.8

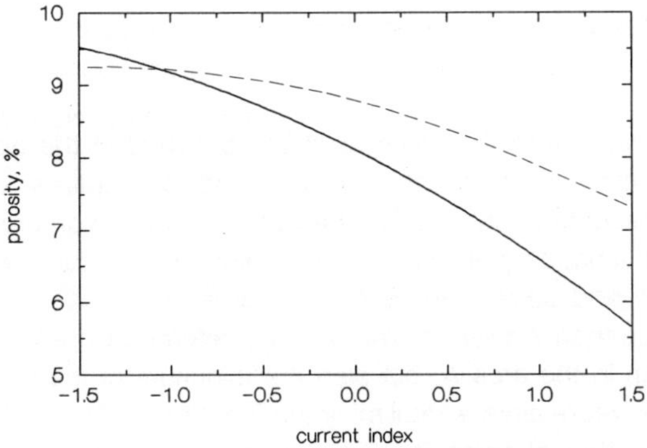

Figure 3. Porosity variation as a function of current index under the condition that the other factors are at zero levels (– – –) and at optimal levels (——) .

In a similar manner, the other properties, wear rate, powder yield and power, can be studied as a function of different variables or their different combinations. An example of this is shown in Fig. 4 presenting the wear rate of the coating as a function of flow rates of H_2 and Ar.

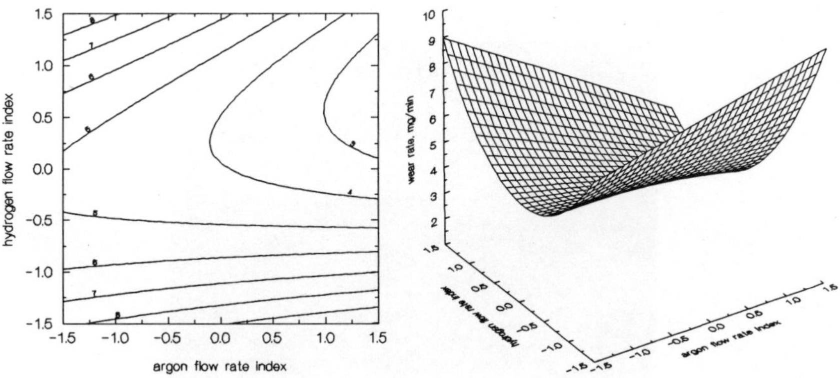

(a) (b)

Figure 4. Wear rate as a function of flow rate indices of argon and hydrogen under the condition that the other factors are at the optimal levels:

a) two dimensional presentation,

b) three dimensional presentation.

As for the optimization by considering several response functions at same time, the minimum porosity, y_1, was searched under the constraints of wear rate y_2 < 7 mg/min, powder yield y_3 > 60 %, and power y_4 < 50 kW. The optimal variables in this case were the same as those obtained in the optimization for porosity only. If more strict constraints are used, the optimal variables may change or even become unobtainable. The optimized operation area can be also determined by overlapping several contour figures. As for the variation of spraying distance and current, there are two areas in Fig. 6, within which the wear rate was less than 7 mg/min. The porosity referring to point P_1 is not only the local minimum in the area A, but also the minimum of the whole test area, while point P_2 only represents a local minimum and the porosity value of which is much higher than that at point P_1. Similarly, the operation area can also be selected in the factorial areas of argon and hydrogen flow rates, as shown in Fig. 7. The value of the minimum porosity raises up as a harsh constraint for wear rate is used, e.g., as the critical wear rate travels from A to B. Determination of the critical constraint value depends on the requirements of practical application. For instance, as the coating is used under a wear condition, the wear rate should be selected as small as possible, e.g., operating at point B; while, under the condition of aggressive corrosion a higher wear rate constraint should be used in order to achieve the lowest porosity, i.e., operating at the point A. In other cases, the constraints can be determined in a similar manner.

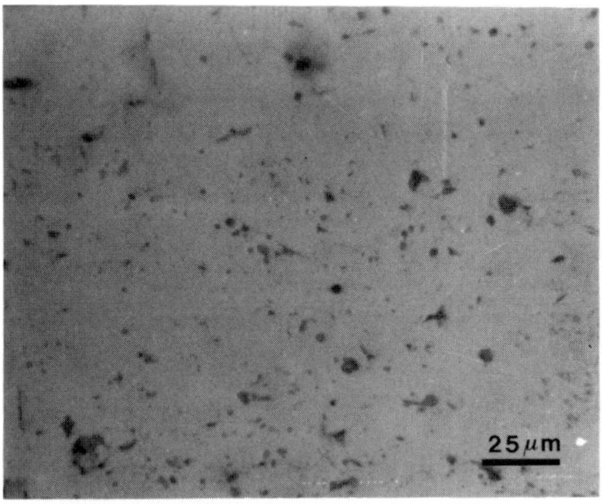

Figure 5. Microstructure of alumina coating sprayed with optimized parameters for minimum porosity.

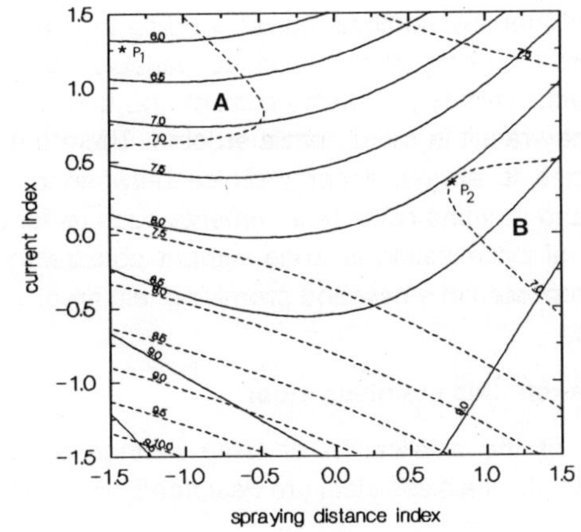

Figure 6. The minimum porosity related not only to factors of spraying distance and current under the condition that the other factors are at the optimal levels, but also affected by the constraint of wear rate:
—— = porosity,
- - - = wear rate.

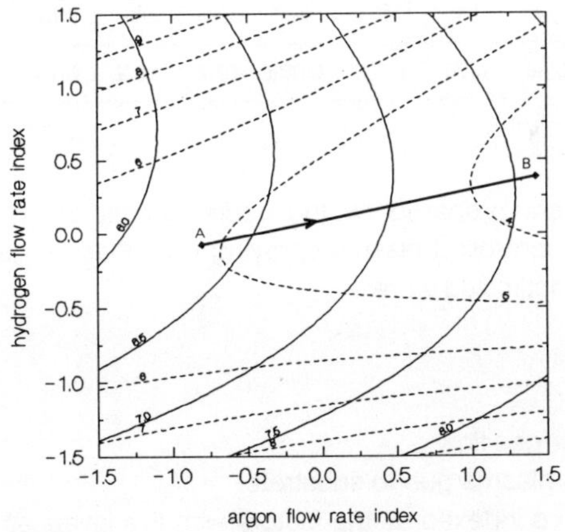

Figure 7. The minimum porosity related to flow rate indices of Ar and H_2 under the condition that the other factors are at the optimal levels, and as affected by the constraint of wear rate:
—— = porosity,
- - - = wear rate.

It was also found that the variables corresponding to minimum porosity and wear rate are located outside of the selected variable variation area. Lower porosity would be achieved by increasing current and lowering argon flow rate. Then, however, it may result in rapid nozzle erosion. Therefore, the selection of the spraying variables is always a compromise between obtaining adequate coating properties and meeting other requirements such as long nozzle life, etc. If the present type of optimization is done without constraints, a further ORO process should be repeated in a new and promising test area.

4. Application to austenitic stainless steel

As an example of the application of ORO method to metallic coatings, experiments on austenitic stainless steel are described in the following.

4.1. Experimental details

The powder used was gas atomized ANVAL 254 SMO with a powder size of 45 - 90 μm. The analysis of the powder is shown in Table 5.

Table 5. Composition of ANVAL 254 SMO plasma spray powder.

Element	C	Si	Mn	P	S	Cr	Ni	Mo	Cu	N
W-%	0.017	0.48	0.76	0.022	0.008	20.2	18.2	6.23	0.82	0.198

The coatings were prepared by the earlier mentioned PLASMA-TECHNIK A-3000S computer controlled plasma spraying unit. Plasma gases were Ar and H_2. Variables to be optimized were:

- z_1, Ar flow
- z_2, H_2 flow
- z_3, current
- z_4, powder carrier gas flow
- z_5, distance from plasma gun to substrate.

The variables were indexed as the factors with five levels as listed in Table 6. The range of the values of different parameters was determined from earlier experiences in spraying stainless steels. The other spraying parameters, given in Table 7, were kept constant in the tests.

The substrates (AISI 4130 steel) were fixed on the lateral surface of a cylinderical specimen holder (200 mm in diameter) which was rotating during spraying at a rate of 120 rpm. The nominal coating thickness was 400 μm.

Table 6. Indices of spraying variables.

x_i	−1.547	−0.1	0.0	1.0	1.547
Argon flow rate [l/min]	25	30	40	50	55
Hydrogen flow rate [l/min]	5	6	8	10	11
Current [A]	423	450	500	550	577
Carrier gas flow rate [l/min]	2.5	2.6	2.8	3.0	3.1
Spraying distance [mm]	112	120	135	150	158

Table 7. Constant parameters in plasma spraying.

Plasma gun	Plasma Technik MB F4
Nozzle/electrode	6 mm
Injector diameter	1.8 mm
Injector angle	90°
Injector distance	8 mm
Powder feed rate	50 g/min
Surface speed of specimen	75 m/min
Traverse speed of gun	2.7 mm/sec

The analyzed coating properties were porosity and oxide content in the coating. The porosity was measured by the water impregnation method and the oxide content by optical microscopy/image analysis from polished cross-sections of the test specimens.

The 27 tests were carried out in a random order to avoid time factors. The most important time factor is the voltage drop, caused by wear of the nozzle in the plasma gun, which results in decrease of gun power. The tests were carried out with a new nozzle and no voltage drop occurred in tests.

4.2. Results

The values of the measured porosity and oxide content as well as the determined coating efficiencies are listed in Appendium 2 (Table 2). The regression equations for porosity, oxide content and coating efficiency were calculated by using Eqs. 6 and 7. *F*-values of each equation and each coefficient were calculated as described in Eqs. 8-15. Finally the equations for porosity (P), oxide content (O), and coating efficiency (E) were obtained as:

$$P = 7.5 + 1.0x_1 - 0.7x_2 - 0.1x_3 + 0.1x_5 - 0.3x_1x_2 - 0.2x_2x_3 - 0.1x_2x_4 - 0.1x_2x_5 - 0.4x_3x_4 + 0.3x_1^2 + 0.1x_2^2 \qquad (20)$$

with $F = 55.1 > F(11, 15)_{\alpha=0.01} = 3.73$

$$O = 16.2 + 1.6x_1 + 1.7x_1x_2 + 1.7x_1x_3 - 1.0x_2x_5 + 1.0x_3x_5 - 1.6x_1^2 + 1.4x_3^2 - 1.3x_4^2 - 1.2x_5^2 \qquad (21)$$

with $F = 5.0 > F(9, 17)_{\alpha=0.01} = 3.68$, and

$$E = 76.5 + 0.9x_1 - 0.9x_3 + 0.7x_1x_2 - 0.6x_2x_3 - 0.8x_2x_5 - 0.6x_4x_5 + 1.3x_1^2 \qquad (22)$$

with $F = (7, 19)_{\alpha=0.01} = 3.77 > F = 3.7 > F = (7, 19)_{\alpha=0.05} = 2.54$

The results of F-test indicate that Eqs. 20 and 21 are at the confidence level of (1 - 0.01) and Eq. 22 at that of (1 - 0.05). The values of porosity calculated by using the determined Eq. 20 and used spraying parameters revealed very good agreement with experimental results, whereas the differences between calculated and measured oxide contents as well as coating efficiency were slightly bigger. It is obvious that the application of these equations at outside of test area is less reliable.

4.3. Optimization

The constraining condition was power, which was limited to 50 kW. The variation range of the factors were determined from -1.5 to 1.5. The optimized levels for factors to get minimum porosity and oxide content are shown in Table 8. Four test runs were carried out to check the reliability of Eqs. 20 and 21. The spray parameters used in the checking sprayings are listed in Table 9. In the first test run the optimized parameters for minimum porosity were used. As Table 10 indicates, the obtained porosity value was higher than the calculated one, but the oxide content was slightly lower than expected. The reason for higher porosity was the too high value of powder carrier gas flow rate which caused the powder to penetrate through the plasma flame. In the second test run, the carrier gas flow rate was decreased as well as the current to get lower oxide content to the coating. The obtained porosity was lower than expected and so also the oxide content. In the third and fourth test sprayings, the original test limits of ORO experiment were exceeded to check how well the equations predict the coating properties outside of test area. The expected porosity values were rather low, but the oxide contents high. As shown in Table 9, the obtained porosity values were higher than expected and those of oxide content were lower. The microstructures in Fig. 8 show clearly the great difference in the coating quality between the coating 28 compared to 30.

Table 8. The optimized levels for spraying variables.

	Porosity		Oxide	
	Index	Value	Index	Value
Argon flow rate [l/min]	–1	30	–1.5	25
Hydrogen flow rate [l/min]	1.5	11	1.5	11
Current [A]	1.5	580	–1	450
Carrier gas flow rate [l/min]	1.5	3.1	1.5	3.1
Spraying distance [mm]	1.5	158	1.5	158

Table 9. The spraying variables used in checking sprayings.

	Spray parameters for checking sprayings							
Test no	28		29		30		31	
	Index	Value	Index	Value	Index	Value	Index	Value
Argon flow rate [l/min]	–1	30	–1	30	–0.5	35	1	50
Hydrogen flow rate [l/min]	1.5	11	1.5	11	3	14	3	14
Current [A]	1.5	580	0	500	2	600	2	600
Carrier gas flow rate [l/min]	1.5	3.1	0	2.8	0	2.8	1	3
Spraying distance [mm]	1.5	155	1.5	155	0	135	0.5	140

Table 10. Comparison of predicted and measured values of optimized porosity and oxide content.

	POROSITY [%]		OXIDE CONTENT [%]	
Test no	Predicted	Measured	Predicted	Measured
28	4.8	6.1	9.4	7.7
29	6.3	4.8	5.2	7.6
30	5.1	6.8	22.6	12.1
31	5.6	6.6	31.6	17.0

(a)

(b)

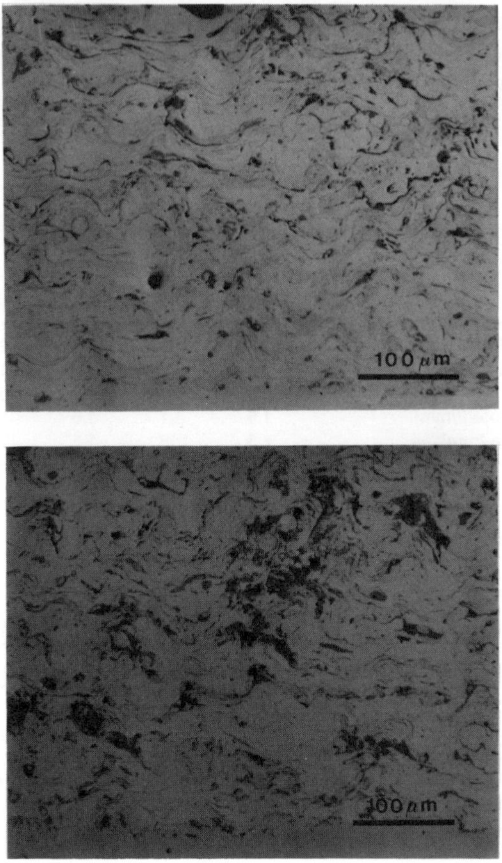

Fig. 8. Microstructures of sprayed coatings:

a) test no 28, optimized parameters for minimum porosity, porosity 6.1 %, oxide content 7.7 %.

b) test no 30, porosity 6.8 %, oxide content 12 %.

4.4. Effect of spraying variables on coating porosity

The argon and hydrogen flow rates are the two major variables effecting to coating porosity as shown in Fig. 9(a) under the conditions that other variables are optimal.

The porosity decreases rapidly as a function of decreasing argon flow rate, when the hydrogen flow rate is low (< -1). At high hydrogen flow rate the effect of argon flow rate is, however smaller. This indicates that at low hydrogen content the heat energy in plasma is inadequate and the particles need more time to melt, i.e., a low argon flow. When the energy of plasma is increased by hydrogen, the optimal point of argon flow is probably shifted towards higher argon flow rates as shown in Fig 9(b), in which the Eq. 20 is applied to hydrogen levels exceeding the levels used in the ORO calculation tests.

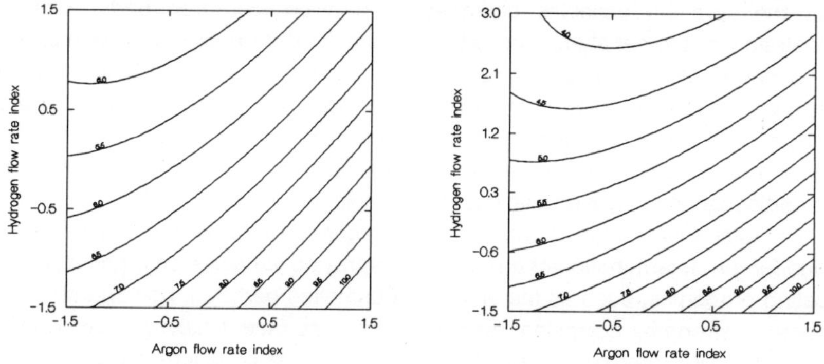

(a) (b)

Figure 9. Porosity variation as a function of flow rate indices of argon and hydrogen:
a) range −1.5 – +1.5,
b) range −3.0 – +1.5.

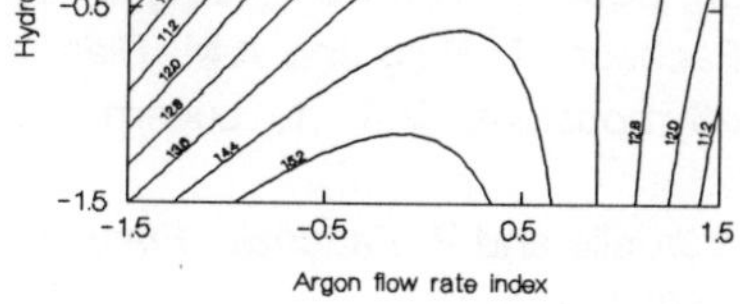

Figure 10. The effect of argon and hydrogen flow rates on the oxide content. The other variables at levels, current = 0, carrier gas = 0, and spray distance = 1.5.

The effects of other variables, i.e., current, carrier gas flow rate and spraying distance are very small as compared to argon and hydrogen flow rates.

The effect of Ar and H_2 flow rate indices on the oxide content according to Eq. 21 is shown in Fig. 10. The minimum value of oxide content is again achieved by low argon flow rates and high hydrogen flow rates.

Conclusion

By using ORO method, the coating properties can be quantitatively predicted and the spraying variables optimized in a given test area under different constraints of the variables. The orthogonal block design allows experiments to be arranged at five levels with reasonable number. It has been found that the dependence of properties on variables differs from one variable variation area to another, regression equation of second order can describe the phenomenon more in detail. Spraying with optimized variables, a coating with the predicted properties can be achieved within an acceptable accuracy.

The total optimization of coating quality requires similar kind of analysis to be done for the determination of the effects of spraying variables on other important coating properties, such as the amount of unmelted particles, bonding strength, cohesive strength, corrosion resistance, and surface finishing. The optimal coating quality may be achieved as a compromise of different properties or under too strong constraints by a multilayer coating in which the various layers have different properties.

Acknowledgement

The develop of ORO method was partially financed by Nordisk Industrifond. The authors want to express the fund their gratitude for this support.

References

1. H. Filmer, J. Hochstrasser and A.R. Nicoll, *Ceramic Bulletin* **69** (1990) p. 1955-8.
2. L. Pawlowski, *Surf. Coatings Tech.* **31** (1987) p. 103-16.
3. V. Costa, S.L. Pacheco, T. Puig and A.M. Dias, in *Proceedings of 2nd Plasma-Technik-Symposium*, **Vol. 1**, Lucerne, Switzerland, 5-7 June 1991, p. 363-72.
4. A. Vardelle, M. Vardelle and P. Fauchais, *Plasma Chemistry and Plasma Processing* **2** (1982) p. 255-91.
5. D.K. Das and R. Sivakumar, *Acta Metall. Mater.* **38** (1990) p. 2187-92.

6. E. Lugscheider, P. Lu, B. Häuser and D. Jäger, *Surf. Coatings Tech.* **32** (1987) p. 215-26.
7. R.B. Heimann, D. Lamy and T. Sopkow, *J. Canadian Ceramic Society* **59** (1990) p. 49-54.
8. T.C.C. Hurng and M.B.C. Quigley, in *Proc. of 2nd Inter. Conf. on Surface Engineering*, Stratford-upon-Avon, UK, 16-18 June 1987, p. 417-21.
9. W.Y. Zhu, *Optimization Design and Application*, (Liaoning Press, Shenyang, China 1981).
10. M.N. Das and N.C. Girl, *Design and Analysis of Experiments* (Wiley Zasterm, Ltd., New Delhi 1979).
11. N.R. Draper and H. Smith, *Applied Regression Analysis* (John Wiley & Sons, Inc., New York, USA 1968).
12. D.A. Wismer and R. Chattergy, *Introduction to Nonlinear Optimization* (Elsevier North-Holland, Inc., New York, USA 1978).
13. K. Schittkowski, NLPQL, A FORTRAN subroutine solving constrained non-linear programming problems, *Annals of Operations Research* **5**, ed. C.L. Monma (1986) p. 485-500.
14. P. Siitonen, S.L. Chen and P. Kettunen, to be published.
15. Standard Designation, *American Society for Testing and Materials* **G65-85**, p. 237.
16. R.A. Johnson and D.W. Wichern, *Applied Multivariate Statistical Analysis* (Prentice-Hall International, Inc., New Jersey 1988).

Orthogonal block design table.

No	x_0	x_1	x_2	x_3	x_4	x_5	x_{12}	x_{13}	x_{14}	x_{15}	x_{23}	x_{24}	x_{25}	x_{34}	x_{35}	x_{45}	x'_1	x'_2	x'_3	x'_4	x'_5
1	1	1	1	1	1	1	1	1	1	1	1	1	1	1	1	1	b	b	b	b	b
2	1	1	1	1	-1	-1	1	1	-1	-1	1	-1	-1	-1	-1	1	b	b	b	b	b
3	1	1	1	-1	1	-1	1	-1	1	-1	-1	1	-1	-1	1	-1	b	b	b	b	b
4	1	1	1	-1	-1	1	1	-1	-1	1	-1	-1	1	1	-1	-1	b	b	b	b	b
5	1	1	-1	1	1	-1	-1	1	1	-1	-1	-1	1	1	-1	-1	b	b	b	b	b
6	1	1	-1	1	-1	1	-1	1	-1	1	-1	1	-1	-1	1	-1	b	b	b	b	b
7	1	1	-1	-1	1	1	-1	-1	1	1	1	-1	-1	-1	-1	1	b	b	b	b	b
8	1	1	-1	-1	-1	-1	-1	-1	-1	-1	1	1	1	1	1	1	b	b	b	b	b
9	1	-1	1	1	1	-1	-1	-1	-1	1	1	1	-1	1	-1	-1	b	b	b	b	b
10	1	-1	1	1	-1	1	-1	-1	1	-1	1	-1	1	-1	1	-1	b	b	b	b	b
11	1	-1	1	-1	1	1	-1	1	-1	-1	-1	1	1	-1	-1	1	b	b	b	b	b
12	1	-1	1	-1	-1	-1	-1	1	1	1	-1	-1	-1	1	1	1	b	b	b	b	b
13	1	-1	-1	1	1	1	1	-1	-1	-1	-1	-1	-1	1	1	1	b	b	b	b	b
14	1	-1	-1	1	-1	-1	1	-1	1	1	-1	1	1	-1	-1	1	b	b	b	b	b
15	1	-1	-1	-1	1	-1	1	1	-1	1	1	-1	1	-1	1	-1	b	b	b	b	b
16	1	-1	-1	-1	-1	1	1	1	1	-1	1	1	-1	1	-1	-1	b	b	b	b	b
17	1	a	0	0	0	0	0	0	0	0	0	0	0	0	0	0	c	d	d	d	d
18	1	-a	0	0	0	0	0	0	0	0	0	0	0	0	0	0	c	d	d	d	d
19	1	0	a	0	0	0	0	0	0	0	0	0	0	0	0	0	d	c	d	d	d
20	1	0	-a	0	0	0	0	0	0	0	0	0	0	0	0	0	d	c	d	d	d
21	1	0	0	a	0	0	0	0	0	0	0	0	0	0	0	0	d	d	c	d	d
22	1	0	0	-a	0	0	0	0	0	0	0	0	0	0	0	0	d	d	c	d	d
23	1	0	0	0	a	0	0	0	0	0	0	0	0	0	0	0	d	d	d	c	d
24	1	0	0	0	-a	0	0	0	0	0	0	0	0	0	0	0	d	d	d	c	d
25	1	0	0	0	0	a	0	0	0	0	0	0	0	0	0	0	d	d	d	d	c
26	1	0	0	0	0	-a	0	0	0	0	0	0	0	0	0	0	d	d	d	d	c
27	1	0	0	0	0	0	0	0	0	0	0	0	0	0	0	0	d	d	d	d	d

a = 1.547, b = 0.23, c = 1.62 and d = -0.77.

Appendium 2

Table 1. Experimental results of the response functions, y_1 = porosity, y_2 = wear rate, y_3 = powder yield and y_4 = power.

No	y_1	y_2	y_3	y_4
1	8.9	7.7	37.9	44.2
2	10.4	14.3	13.3	37.2
3	8.3	6.5	57.0	42.2
4	9.2	10.1	41.8	35.1
5	9.2	9.8	30.7	34.5
6	11.6	19.2	10.6	27.9
7	8.9	8.8	57.2	32.5
8	10.5	13.3	35.6	26.1
9	8.3	4.8	37.7	43.0
10	8.7	8.4	26.9	36.0
11	7.0	7.0	64.7	42.9
12	8.5	7.9	49.9	34.5
13	10.7	11.0	33.9	32.7
14	10.4	16.4	17.9	26.5
15	8.5	8.8	50.8	32.0
16	10.1	11.0	35.8	25.7
17	10.4	8.7	40.1	34.2
18	8.5	6.6	41.7	33.9
19	7.2	6.5	49.4	41.0
20	9.4	10.6	34.1	27.0
21	9.2	8.4	24.4	34.0
22	8.3	7.6	65.1	33.4
23	8.4	7.3	50.4	38.0
24	10.6	15.1	20.9	27.0
25	9.5	8.6	35.2	33.0
26	9.5	8.0	36.0	32.9
27	8.7	8.0	39.9	33.0

Table 2. Porosity, oxide content and coating efficiency.

Test no	Porosity [%]	Oxide [%]	Efficiency [%]
1	7.2	16.5	75
2	8.2	21.1	79
3	8.5	16.2	81
4	8.0	11.6	81
5	9.3	11.9	79
6	10.5	14.1	78
7	10.2	10.6	78
8	9.2	14.5	78
9	5.6	8.1	77
10	6.7	8.7	73
11	6.8	7.8	77
12	6.6	16.0	76
13	7.0	11.7	79
14	7.3	10.5	75
15	7.5	15.9	77
16	7.1	16.3	79
17	9.4	17.2	80
18	7.0	9.3	80
19	6.7	16.4	76
20	8.7	15.2	76
21	7.3	24.2	73
22	7.8	16.9	79
23	7.4	15.0	76
24	7.2	13.1	80
25	7.7	15.8	78
26	7.2	12.4	77
27	8.0	13.9	73

Chapter 4

AN ANALYTICAL APPROACH TO PLASMA SPRAYING

Shrikant V. Joshi* and R. Sivakumar+
* **Defence Metallurgical Research Laboratory**
P.O.Kanchanbagh, Hyderabad-500 258.
\+ Sree Chitra Thirunal Institute of Medical Sciences
and Technology, Trivandrum, India

ABSTRACT

An analytical approach to plasma spraying is recognized to have a significant role to play in further development of this promising surface modification tool. A theoretical prediction model that can potentially assist adoption of such an approach is presented. The model accounts for the various factors that influence plasma-particle heat and momentum transfer and can realistically estimate the in-flight velocity, temperature and size of a particle during plasma spraying. The mathematical analysis associated with the model is detailed. The validity of the model as well as its utility in both the plasma and high velocity oxy-fuel spray techniques are also highlighted through some example calculations performed under typical spraying conditions.

1. INTRODUCTION

Plasma-sprayed coatings have been gaining increasing acceptance in recent years to enhance the service life of critical components which perform under adverse operating conditions. These coatings serve as the first line of defence against attack by an aggressive environment and provide protection against the commonly encountered surface degradation mechanisms such as corrosion, oxidation, wear or failure under an excessive heat load [1,2]. The formation of a judiciously chosen high-performance coating on a component fabricated using an appropriate high mechanical strength metal/alloy offers an attractive means of meeting both the bulk and surface property requirements of nearly every imagined application. This surface modification concept can play a vital role in curtailing down times and avoiding frequent damage of expensive components.

Recognition of the tremendous potential of coatings technology has led to rapid advancement in this field during the past decade. A variety of "dedicated" spray powders for coating applications are now available [3] enabling surface-coated components to be used in a wide range of operating conditions [4]. Even more significantly, numerous competing thermal spray techniques for forming the protective coatings have been developed [5,6]. While the plasma spraying method is particularly well-accepted [7,8], the relatively new high velocity oxy-fuel (HVOF) spraying techniques are also being increasingly employed [9,10]. It has been found that HVOF methods can even yield superior coatings than conventional plasma deposition for certain categories of coatings which involve spraying of metals and low melting alloys [11].

Regardless of the obvious promise of the two above mentioned coating techniques, their full potential can be realized only by ensuring that high quality protective coatings are applied reproducibly and economically. The quality of a thermally sprayed coating, especially as regards to porosity and adhesion strength, depends primarily upon the extent of particle melting and the particle velocity at the moment of impact with the substrate. In order to obtain dense and well-bonded coatings, it is required that the powder particles be completely molten when they impinge on the substrate and that the particle velocity at impact be sufficiently high. However, *in situ* monitoring of the particle parameters during plasma and HVOF spraying is both expensive and difficult. As such, it is clear that an analytical approach to the coating processes can play a significant role in further development of the spray techniques.

The particle temperature and velocity at any instant during its traverse through a plasma flame is governed by the gas-solid heat and momentum transfer processes between the plasma gases and the injected powder particles. Therefore, reliable prediction of the particle parameters during spraying is largely dependent upon the ability to realistically model the above transport phenomena by accounting for the various factors that can influence gas-solid transport rates under typical spraying conditions. A theoretical model, considering all factors relevant to a plasma spray process, has recently been developed at this laboratory for estimation of the velocity, temperature and size of a particle at any given axial distance from the plasma gun nozzle [12]. This model enables accurate predictions and can be a valuable tool for adopting an analytical approach to plasma and HVOF spraying [12,13]. The present article discusses the various phenomena that influence plasma-particle heat and momentum transfer, includes details of the mathematical analysis associated with the model and provides certain illustrative results to highlight the validity as well as the utility of the model.

2. FACTORS INFLUENCING PLASMA-PARTICLE TRANSPORT PROCESSES

The modelling of transport processes during plasma spraying presents certain unique problems which are alien to the normally encountered non-isothermal gas-solid systems. For instance, the difference between the particle surface and the bulk plasma gas temperatures can be as high as 10,000 K and this makes the choice of a single temperature for transport property evaluation difficult [14]. The problem becomes particularly pronounced owing to the peaks in the plasma property-temperature relationship arising due to ionization and/or dissociation of the plasma gas. Under such conditions, it has been shown that the use of an integral mean value of the thermal conductivity is most appropriate for heat transfer calculations [14].

Another phenomena that can considerably influence the gas-solid transport rates during spraying owes its significance to the typically fine particle size of spray powders. These powders usually consist of particles less than 50-100 μm in size so as to facilitate complete melting. In such circumstances, the magnitude of the gas mean free path is comparable with the particle diameter and the continuum transport mechanisms fail to completely describe the transport processes [15]. A measure of the departure from the laws of continuum mechanics (known as the Knudsen effect) is provided by a dimensionless parameter called the Knudsen number, which is defined as [16]

$$Kn = \lambda / d_p \tag{1}$$

where λ is the gas mean free path and d_p the particle diameter. It has now been established that the Knudsen effect can dominate the heat, mass and momentum transfer processes under plasma conditions by substantially decreasing the gas-solid transport rates [17-19].

In addition to the above factors, the internal heat conduction in particles (even as small as 50 μm in diameter) is also known to play a significant role in the overall plasma-particle heat transfer process. In the case of ceramic powders, which have a low thermal conductivity and are extensively used for a variety of thermal barrier applications, it has been shown that the particle surface can even reach boiling point while the particle centre is still solid [14]. Hence, neglecting the contribution of intra-particle heat transfer may not always be justified. It is also pertinent to reiterate here that, rather than melting of merely the particle surface, it is the complete melting of the particles to the core that is desirable for obtaining dense coatings.

3. MODEL DETAILS

During the past decade, numerous mathematical models were developed for the prediction of in-flight particle parameters during plasma spraying. An exhaustive review of most of these plasma-particle modelling efforts is available in Ref. [20]. However, the applicability of the existing models to any actual plasma spraying operation is either inappropriate or severely restricted owing to certain simplifying assumptions which completely ignored one or more of the various important factors influencing plasma-particle transport processes. For instance, certain studies did not account for the steep temperature gradients [21], a few neglected the influence of intra-particle heat transfer [22,23], some chose to ignore the Knudsen non-continuum effects [14,24] and yet others considered a constant plasma temperature [17,18] which is clearly unrealistic considering the rapid temperature decay with axial distance in an actual plasma flame.

Motivated by the existing need of a capability to realistically predict the in-flight particle parameters during plasma spraying, a comprehensive one-dimensional model was recently developed by the authors. This model, briefly described below, simultaneously considers internal heat conduction in particles, accounts for the steep temperature gradients and incorporates the influence of the Knudsen effect. Furthermore, this model visualizes a realistic case wherein the traversing particle sees a varying flame temperature and accounts for the corresponding change in all thermophysical gas properties. Although initially developed, and used, to investigate the plasma spray process [12], this model is by no means restrictive and has already been extended to be applicable to HVOF spraying [13].

3.1. Analysis

The problem formulation presented in this section is based on the assumptions that the flow is one dimensional and that the particle loading is low enough so as not to influence the temperature and velocity profiles in the plasma flame. It may be pointed out that the latter assumption can be eliminated by using the Particle-Source-In-Cell method [25] to incorporate particle loading [26]. However, over the range of powder feed rates typically employed in plasma spraying, the particle temperature and velocity are found to remain virtually unchanged [27], thereby indicating that the plasma velocity and temperature fields are not significantly affected by particle loading of this magnitude. The above assumption may, therefore, be retained to simplify and speed-up computations without compromising on the accuracy of its predictions. The heat and momentum transfer processes between the plasma gases and a spherical particle of diameter d_p can then be described as outlined below.

3.1.1. Gas-Particle Heat Transfer

The heating up of a spherical particle injected into a plasma flame is governed by the transient heat conduction equation

$$\rho_p c_p \frac{\partial T}{\partial t} = \frac{1}{r^2} \frac{\partial}{\partial r} (r^2 k_p \frac{\partial T}{\partial r}) \qquad (2)$$

where t is the time coordinate, r is the radial coordinate within the particle and T the corresponding temperature. ρ_p, c_p and k_p denote the density, specific heat and thermal conductivity, respectively, of the particle material.

If T_{po} is the initial uniform particle temperature at the time of injection, the initial condition for solving the above governing equation is

$$t=0, \text{ all } r<d_p/2, \ T(r,t)=T_{po} \qquad (3)$$

The two boundary conditions at the particle surface and the particle core may be written as

$$t>0, \ (\partial T/\partial r)_{r=0} = 0 \qquad (4)$$

$$t>0, \ [k_p(\partial T/\partial r)]_{r=d_p/2} = h \ (T_f-T_{ps}) \qquad (5)$$

where T_f is the bulk combustion gas temperature and T_{ps} is the particle surface temperature.

The heat transfer coefficient h in Eq. (5) can be calculated from the Nusselt number Nu estimated from the semi-empirical Ranz-Marshall correlation [28]

$$Nu = \frac{h \ d_p}{\bar{k}_f} = 2.0 + 0.514 \ Re^{0.5} \qquad (6)$$

In the above equation, $\bar{k}_f$ is the integral mean thermal conductivity, defined exclusively to account for the steep temperature gradients in the thermal boundary layer [14], and is given by

$$\overline{k}_f = \frac{1}{T_f - T_{ps}} \int_{T_{ps}}^{T_f} k_f \; dT \tag{7}$$

It may be noted that the above expression is valid even when the difference between the particle surface and bulk gas temperatures is small or when the thermal conductivity does not vary appreciably over the temperature range in question.

The particle Reynolds number Re required to calculate the heat transfer coefficient from Eq. (6) is defined by

$$Re = d_p \, |U_p - U_f| \; \rho_f / \mu_f \tag{8}$$

where U_p and U_f are the particle and the bulk plasma gas velocities, respectively, while $_f$ and μ_f are the integral mean values of the gas density and gas viscosity determined from expressions similar to Eq. (7).

As mentioned previously, the Knudsen discontinuum effects can significantly influence the gas-particle transport phenomena during plasma spraying. The Knudsen effect on heat transfer manifests itself in the form of a temperature jump at the particle surface [29] whereby the gas temperature in the immediate particle vicinity T_{pv} is higher than the particle surface temperature T_{ps}. This leads to a reduced heat flux, as the effective driving force for heat transfer $(T_f - T_{pv})$ in the case of a temperature jump is lower than $(T_f - T_{ps})$, the driving force which would exist in the absence of discontinuum effects. Chen and Pfender [17] have shown that the ratio of the two driving forces, which effectively serves as a correction factor $\emptyset$ for heat flux calculations, can be estimated from the expression

$$\emptyset = \frac{T_f - T_{pv}}{T_f - T_{ps}} = \frac{1}{1 + 4\left(\frac{2-\theta}{\theta}\right)\left(\frac{\Gamma_{ps}}{1+\Gamma_{ps}}\right)\left(\frac{Kn^*}{Pr_{ps}}\right)} \tag{9}$$

In the above equation, θ is the thermal accommodation coefficient, Γ the specific heat ratio and Pr_{ps} the Prandtl number $(Pr = c_p \mu / k)$ evaluated at the particle surface temperature. Kn^* is the modified Knudsen number defined as [17]

$$Kn^* = \frac{2\,Pr_{ps}}{\rho_{ps}\,v_{ps}\,d_p}\,\frac{k_{f,t}}{c_{p,t}} \tag{10}$$

where $k_{f,t}$ and $c_{p,t}$ are the integral mean values of the thermal conductivity and specific heat of the plasma gases over the range of the temperature jump T_{ps} to T_{pv}. The mean molecular speed v_{ps} is dependent upon the average molecular weight M of the gas mixture as well as the particle surface temperature T_{ps} and is given by

$$v = (8RT/M\pi)^{1/2} \tag{11}$$

where R is the gas constant.

It is to be noted that, when the Knudsen effect is negligible and $Kn^* \longrightarrow 0$ (as in the case of a large particle), the correction factor ∅ estimated from Eq. (9) approaches a value of unity and T_{pv} approaches T_{ps}. However, the particle size is necessarily small (<100 μm) in plasma spraying. Under such circumstances, consideration of the Knudsen effect becomes almost mandatory for a model to enable realistic predictions. It should further be pointed out that T_{pv} is not an *a priori* known quantity and should be determined from Eq. (9) by an iterative procedure as indicated elsewhere [30].

When the particle surface eventually reaches the melting temperature, a heat balance at the moving solid-liquid interface must account for the latent heat associated with melting. Thus,

$$\Delta H_m\,\rho_{p,s}\,\frac{dr_m}{dt} = k_{p,l}\,\frac{\partial T}{\partial r}\bigg|_{r=r_{m,l}} \tag{12}$$

where ΔH_m is the latent heat of fusion and r_m is the radial position of the melting front, with subscripts s and l designating the solid and molten regions in the particle.

If the particle size is fine enough (as in the case of most spray-grade ceramic powders) or its boiling point low (as in the case of metallic powders used for spraying applications), the particle can even vaporize during spraying. The change in particle size due to vaporization can be described using a relationship similar to Eq. (12) as

$$\Delta H_v\,\rho_{p,s}\,\frac{dr_b}{dt} = h(T_f - T_{bp}) - k_{p,l}\,\frac{\partial T}{\partial r}\bigg|_{r=r_b} \tag{13}$$

where H_v is the latent heat of vaporization and r_b the radial position of the boiling front. It is tacitly assumed that, once the particle begins to evaporate, its surface temperature remains constant at the boiling point T_{bp} until a stage is reached when the particle begins to cool.

3.1.2. Gas-Particle Momentum Transfer

The governing equation for gas-particle momentum transfer associated with plasma spraying can be written as

$$\frac{dU_p}{dt} = \frac{3}{4} \frac{C_D}{d_p} \frac{\rho_f}{\rho_p} (U_f - U_p) |U_f - U_p| \qquad (14)$$

The drag coefficient C_D in the above equation is a function of the particle Reynolds number and, depending upon the Re value, can be estimated from the most appropriate of the various empirical correlations available in the literature [12].

Chen and Pfender [18] have shown that the Knudsen effect greatly influences the momentum transfer too and the drag coefficient is corrected as follows to account for this effect:

$$C_D = C_{D,C}\, \phi^{0.45} \qquad (15)$$

In the above equation, $C_{D,C}$ represents the value of the drag coefficient estimated using any one of the correlations mentioned above while ϕ is the correction factor given by Eq. (9).

The initial condition for solving the governing equation for momentum transfer is

$$t = 0, \quad U_p = U_{po} \qquad (16)$$

where U_{po} is the particle injection velocity. As the axial location of the powder injection port ($z=z_o$) is known and since $dz/dt=U_p$, the axial position of an injected particle at any given time t can be readily computed.

3.1.3. Numerical Solution Procedure

Calculations of the in-flight velocity, temperature and size of an injected particle during plasma spraying can be made based on the transport equations formulated above. As previously mentioned, the intra-particle heat transfer also needs to be considered to enable accurate predictions. For the purpose of computation, the resulting set of equations can be discretized using a finite difference scheme and a variable grid method [31] may be adopted for solving the moving boundary problem arising when the particle begins to melt and/or evaporate. Further details of the solution procedure are provided elsewhere [12].

4. ILLUSTRATIVE RESULTS AND DISCUSSION

The above model was initially developed, and used, to investigate the plasma spray process. Calculations of the velocity, temperature and size of a particle injected into a 5:1 Ar/H_2 plasma were made using the experimentally determined data from Vardelle et al. [23] for velocity and temperature fields in the plasma flame. The significance of the various factors considered in this model on plasma-particle transport processes was amply revealed by the computation results [12]. The adaptability of the model to the HVOF spray process has also been shown to be readily feasible [13]. While the interested reader is referred to the two above mentioned papers for detailed theoretical studies on particle behaviour during plasma and HVOF spraying, for reasons of brevity it is intended to highlight herein only the validity of the present model and the utility of adopting an analytical approach, based on this model, to assist plasma and HVOF spraying.

4.1. Model Validation

The experimental verification of models developed for plasma-particle systems is usually extremely difficult owing to the problems associated with *in situ* measurement of particle parameters. A convenient method to check the estimated particle evaporation is to plasma spray powders of known starting particle diameter into water from a fixed distance and to compare the experimentally determined final particle size with model predictions. Similar experiments with ceramic powders have already revealed good agreement between experimental data and predictions of the above model [12]. Although the above experiment is informative to assess the accuracy of the model predictions, it is of more relevance to plasma powder synthesis processes than it is to spraying operations and, therefore, its results are not reproduced here.

Considering the importance of particle velocity in spraying applications, it was further decided to compare the predicted particle velocity evolution with the experimental data of Vardelle et al. [23] from whose study the plasma flame velocity and temperature fields for the computations were also taken. Figure 1 confirms that the model predictions very closely match the experimental data. It also reveals that the estimates of this model are a marked improvement over the predictions of other models [23,24]. The enhanced accuracy of the present model is attributed primarily to the inclusion of the steep temperature gradient and non-continuum effects, the latter being overlooked in both the other models. Unfortunately, reliable particle temperature data is not available in the literature owing to the difficult diagnostics involved and precludes similar comparison.

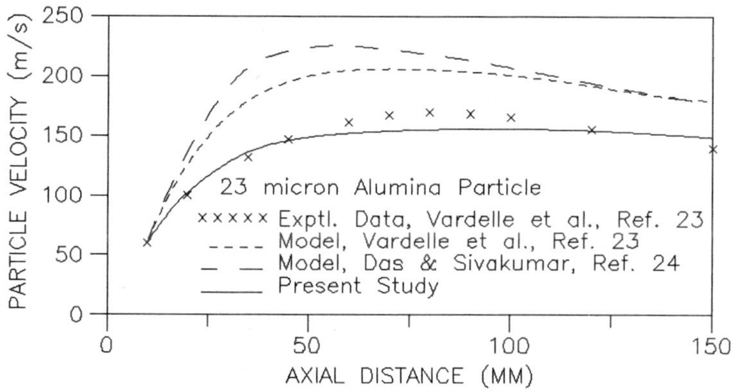

Figure 1. Validation of the present model with experimental data [23] and comparison with predictions of existing models [23,24].

4.2. Utility of Prediction Method

As mentioned earlier, the most important particle parameters from the standpoint of coating quality are the extent of particle melting and its velocity at impact. In this regard, the model provides a very convenient tool to study particle heat-up and acceleration during plasma spraying. The evolution of the particle centre temperature as a function of time is depicted in Fig. 2 for zirconia particles of three different sizes. This figure also illustrates the importance of the Knudsen non-continuum effect on plasma-particle heat transfer. The Knudsen effect is found to become progressively more marked as the particle size is reduced. This conforms to expectations, as the Knudsen number, which provides a measure of the significance of the non-continuum effects, varies inversely as the particle diameter [16].

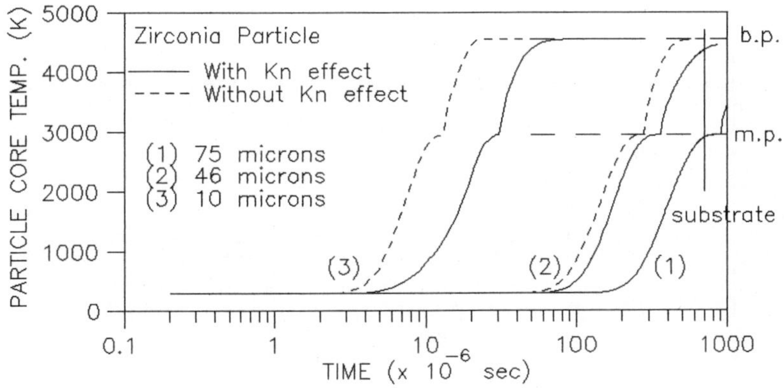

Figure 2. Effect of particle size on particle heat-up during plasma spraying.

As illustrated in Fig. 2, the model further reveals that, for a 75 µm zirconia particle, the particle centre does not attain the melting point by the time it impacts the substrate. Therefore, an upper size limit of about 50 µm is deemed more appropriate to ensure complete melting of zirconia powder during plasma spraying. The model allows similar determination of the maximum particle size that will allow complete melting of any given powder material in the plasma flame. Predictions such as these can be of invaluable assistance in choosing powders of a suitable size range for plasma spraying.

Figure 3 shows the velocities of zirconia particles of three different sizes as a function of axial position. It is apt to mention here that the drag coefficient C_D has to be corrected as indicated in Eq. (15) to account for the significant Knudsen effect. Furthermore, as pointed out earlier, ignoring the Knudsen effect during heat transfer calculations leads to an overestimation of the heat flux and, therefore, to a lower predicted particle size by virtue of an exaggerated rate of evaporation [12]. As smaller particles are accelerated much faster than the larger ones by the plasma gas flow, a higher maximum particle velocity is predicted if the reduction in the heat flux due to Knudsen effect is ignored. The results shown in Fig. 3 have been obtained by taking the above factors into account and these reveal that, as in the case of particle heat-up, the particle size significantly influences the magnitude of the particle velocity at impact with the substrate.

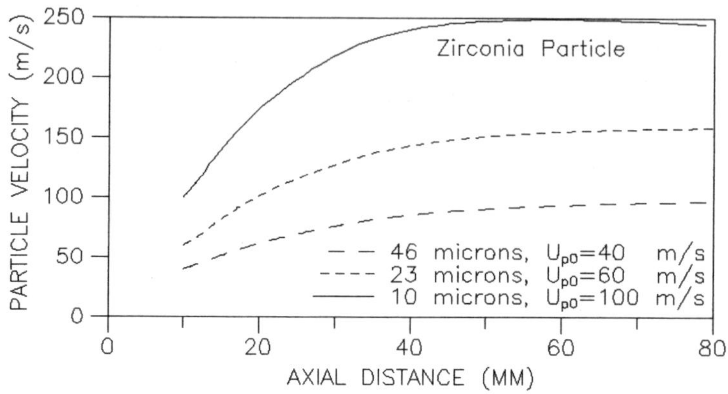

Figure 3. Influence of particle size on particle acceleration during plasma spraying.

It may be pointed out that U_{po} in Fig. 3 is the particle injection velocity which, for a given carrier gas flow rate and for a fixed powder feed rate, increases with decreasing particle size. Incidentally, the particle injection velocity as well as the location of the powder injection port have a considerable influence on particle behaviour in a plasma flame - while both have a bearing on the particle residence time in the flame region, the latter also determines the highest temperature that the particle is exposed to.

The particle heat-up and acceleration estimates like those illustrated in Figs. 2 and 3 can be used in conjunction to optimize the spray distance such that the particle velocity at impact is maximized, without compromising on the extent of particle melting due to cooling in the plasma flame tail region.

4.3. Adaptability of the Model to HVOF Spraying

The use of the above model is not restricted to plasma spraying alone and the model can be easily modified to study particle heating and acceleration during HVOF spraying. Some illustrative results of particle behaviour during HVOF spraying obtained by using the present model have already been published elsewhere [13]. The model can be gainfully employed to assess the relative advantages/shortcomings of plasma and HVOF methods for spraying a given powder.

An HVOF spraying system is known to provide a gas

velocity of around 1200 m/s and flame temperatures in the 3000-3500 K range [32]. These values differ greatly from those prevailing in a plasma flame. It has been shown that the low temperatures typical of an oxy-fuel combustion flame preclude complete melting of most ceramic powders [13], which is in stark contrast to even coarser ceramic particles undergoing some evaporation during plasma spraying [12]. However, the high velocity combustion flame can accelerate the injected particles considerably. Figure 4 shows the velocity evolution of 25 μm particles of three different powder materials. It is observed that velocity evolution of the particles depends strongly on the powder material, with the lighter alumina particles being accelerated much more

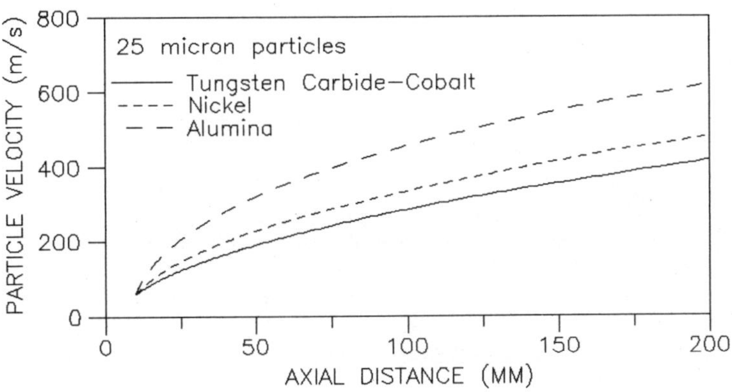

Figure 4. Evolution of particle velocities of different powder materials during HVOF spraying.

than WC-12%Co particles of the same size, which are almost four times as heavy. Comparing Figs. 3 and 4, it is also evident that the particle velocity at impact is much greater in the case of HVOF spraying than in the case of plasma spraying. This provides the HVOF spray process a distinct advantage for the development of dense, well-bonded coatings using metallic/alloy powders that possess a low melting point and can, therefore, be easily molten in the combustion flame.

An additional advantage of the HVOF technique is its relative insensitivity to the particle injection parameters such as the injection velocity. For instance, as shown in Fig. 5, a substantial change in the particle injection velocity has only a marginal effect on the temperature evolution of the particle. Likewise, its influence on the particle velocity at impact is also negligible [13]. This may be attributed to the fact that a HVOF system provides nearly uniform heat input and acceleration over much of the spraying distance [32] and, therefore, a small variation in the particle dwell time brought about by changing the injection parameters has little effect. In contrast, the velocity and temperature fields in the plasma flame decay rapidly with axial distance. As such, the quality of plasma sprayed coatings is extremely sensitive to the injection velocity as well as the axial location of the powder injection port [24] since both considerably affect the plasma-particle transport processes.

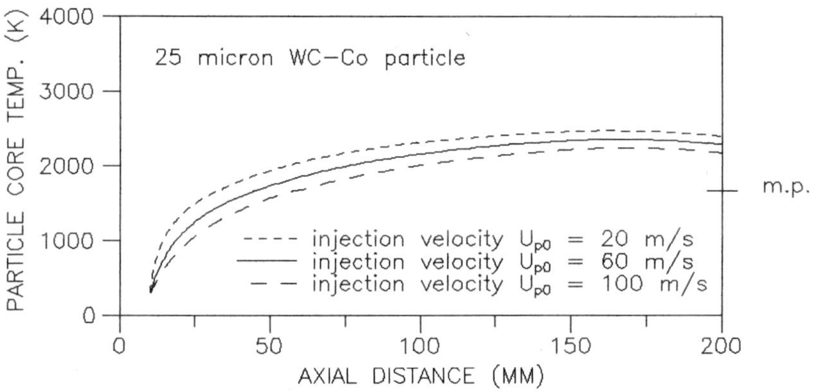

Figure 5. Effect of particle injection velocity on particle heat-up during HVOF spraying.

5. CONCLUSIONS

A theoretical model is presented for particle dynamics and heat transfer under plasma conditions. The model enables an accurate prediction of the velocity, temperature and size of a particle during its traverse through a plasma flame. Such a model constitutes an important step in the overall effort aimed at enabling an analytical approach to the plasma spraying process. The analysis of plasma-particle momentum and heat transfer simultaneously considers internal heat conduction in particles, accounts for the steep temperature gradients and incorporates the influence of the Knudsen discontinuum effects on both heat and momentum transfer.

The predictions of the model are examined through some example calculations performed under conditions generally encountered in typical plasma processing operations. The results illustrate the accuracy of the model estimates. Such a quantitative model can clearly bring out the relative importance of the large number of spraying variables, without which optimisation is an arduous task. Furthermore, it is found that the model can be easily extended to be applicable to HVOF spray processes and provides a convenient means to assess the relative advatages/shortcomings of the two coating methods. Briefly, the calculations lead to the following conclusions:

1. Comparison with experimental data reveals that predictions of the present model are accurate. The model, which is reasonably simple and quite easy to use, can potentially be a very useful tool for the model-assisted development of various plasma-particle systems.

2. The Knudsen non-continuum effect on gas-particle heat transfer, during both plasma and HVOF spraying, can be significant. The importance of this effect becomes increasingly pronounced as the particle size is reduced.

3. The influence of the Knudsen effect on gas-particle momentum transfer is also considerable in both cases and it should be accounted for to obtain accurate particle velocity predictions.

4. The particle heat-up and acceleration during spraying are greatly influenced by the particle size. Therefore, the particle diameter is an important parameter in determining the extent of particle melting as well as the particle velocity at impact and, thereby, the coating porosity.

5. The HVOF techniques can accelerate the powder particles much more rapidly than plasma spraying can and this can be a distinct advantage in the development of dense, well-bonded coatings.

6. Plasma spraying is much more versatile than HVOF spraying - while the latter provides a distinct advantage for spraying metallic/alloy powders that possess a low melting point, the low inherent temperatures provided by an oxy-fuel combustion flame are inadequate for complete melting of ceramic powders.

REFERENCES

1. R.W.Smith and R.Novak, Pow. Metall. Intl., **23** (1991) 147.

2. E.Lugscheider, H.Eschnauer, U.Muller and Th.Weber, Pow. Metall. Intl., **23** (1991) 33.

3. K.G.Budinski, "Surface Engineering for Wear Resistance" (Prentice Hall Inc., New Jersey, 1988).

4. R.Sivakumar and S.V.Joshi, Trans. Indian Ceramic Society, **50** (1991) 1.

5. T.N.Rhys-Jones, Surf.Coat.Tech., **42** (1990) 1.

6. I.Kvernes, M.Espeland and O.Norholm, Scand. J. Metall., **17** (1988) 8.

7. J.H.Zaat, Ann. Rev. Mater. Sci., **13** (1983) 271.

8. I.A.Fisher, Intl. Metall. Rev., **17** (1972) 115.

9. H.Kreye, Proc. Natl. Thermal Spray Conf., Cincinnati, October 1988 (ASM International, Metals Park, Ohio, 1989).

10. C.Verport and G.Loewe, in "Advances in Surface Treatments," Vol. 5, ed. A.Niku-Lari (Pergamon, New York, 1987).

11. K.V.Rao, D.A.Somerville and D.A.Lee, Proc. 11th Intl. Thermal Spray Conf., Montreal, September 1986 (Welding Institute of Canada, 1986).

12. S.V.Joshi and R.Sivakumar, Mat. Sci. Tech., in press (1992).

13. S.V.Joshi and R.Sivakumar, Surf. Coat. Tech., **50** (1991) 67.

14. E.Bourdin, P.Fauchais and M.Boulos, Int. J. Heat Mass Transfer, **26** (1983) 567.

15. B.Waldie, Chem. Engng., London, **259** (1972) 188.

16. S.A.Schaaf and P.L.Chambre, in "Fundamentals of Gas Dynamics," ed. H.W.Emmons (Princeton University Press, Prinecton, New Jersey, 1958).

17. Xi Chen and E.Pfender, Plasma Chem. Plasma Process., **3** (1983) 97.

18. Xi Chen and E.Pfender, Plasma Chem. Plasma Process., **3** (1983) 351.

19. S.V.Joshi, J.Y.Park, P.R.Taylor and L.S.Richardson, Int. J. Heat Mass Transfer, **29** (1986) 1565.

20. S.V.Joshi, Ph.D. Thesis, University of Idaho (1987).

21. J.K.Fiszdon, Int. J. Heat Mass Transfer, **22** (1979) 749.

22. Xi Chen and E.Pfender, Plasma Chem. Plasma Process, **2** (1982) 185.

23. M.Vardelle, A.Vardelle, P.Fauchais and M.I.Boulos, AIChE J., **29** (1983) 236.

24. D.K.Das and R.Sivakumar, Acta Metall., **38** (1990) 2187.

25. C.T.Crowe, M.P.Sharma and D.E.Stock, J. Fluids Engng., **99** (1977) 325.

26. J.W.Mckelliget and N.El-Kaddah, "Plasma Processing and Synthesis of Materials III", MRS Symp. Series Vol.190, ed. D.Apelian and J.Szekely, (MRS, Pittsburgh, 1990).

27. J.R.Fincke and W.D.Swank, "Plasma Processing and Synthesis of Materials III," MRS Symp. Series Vol.190, ed. D.Apelian and J.Szekely (MRS, Pittsburgh, 1990).

28. W.E.Ranz and W.R.Marshall, Chem. Engng. Prog., **48** (1952) 141.

29. E.H.Kennard, "Kinetic Theory of Gases," (McGraw-Hill, New York, 1938).

30. S.V.Joshi, J.Y.Park, P.R.Taylor and L.S.Richardson, Plasma Chem Plasma Process., **6** (1986) 281.

31. T.Yoshida and K.Akashi, J. Appl. Phys., **48** (1977) 2252.

32. D.L.Mason and K.Rao, Proc. 2nd Natl. Conf. on Thermal Spraying, Long Beach, 1984.

Chapter 5

OPTIMISING THE APS-PROCESS PARAMETERS FOR NEW NI-HARDFACING ALLOYS USING A MATHEMATICAL MODEL

Univ.-Prof. Dr.techn. E. Lugscheider*, Dr.-Ing. B.C. Oberländer** and Dipl.-Ing. A. Sabour Rouhaghdam*

* Material Science Institute, RWTH Aachen, Templergraben 55, D-5100 Aachen, Germany

** now: Metallurgy and Materials Engineering, Kath. Universiteit Leuven, de Croylaan 2, B-3001 Leuven, Belgium

Abstract

A brief introduction into the principles of two-level factor analysis is given. It is then shown how, with the help of statistics, known experimental process-parameters, and resulting target parameters of the coating, the two-level factor analysis (a statistical test plan) can be constructed. It reduces the number of test experiments necessary to optimise process-parameters in the plasma spray process for achieving the desired microstructural morphologies with a given coating material (powder). The method is used to optimise the atmospheric pressure plasma-spray process (APS-process) for newly developed τ-boride Ni-hardfacing alloys. Plasmasprayed coatings made of powders of this new hardfacing alloys can be used to protect exposed surfaces of technical components against wear in corrosive environments. These alloys possess a good abrasion resistance, and due to the chromium content a good resistance against alkaline solutions, nitric and sulphuric acids, and against oxidation at higher temperatures.

Introduction

The coating engineer is often faced with the task of developing coating procedures, e.g. plasma-spray procedures, for new coating materials according to the coating characteristics required. The plasma-spray process, however, involves a complex system of variables, of experimental parameters /1/. The choice and control of the multitude of experimental parameters affect the coating qualities. The coating qualities obtained in the process, can be quantified by easily measurable properties, the so called target quantities such as porosity, coating thickness, surface roughness etc..

The knowledge of the interaction of spraying parameters and coating characteristics is an essential factor for optimising coating qualities. The influences of the experimental parameters on the coating qualities achieved can be examined either by a strictly systematic method, requiring a large number of experiments or by a statistical test plan. With the first method only one parameter is varied per test series and its influence on the target quantity determined whilst the remaining experimental parameters are kept constant. In this way the individual effects can be determined, but not the interactions between the parameters. This disadvantage can be avoided using a statistical test plan known as the two level factor analysis, with the added benefit that the objective can be achieved with a much lower number of experiments.

Based on known experimental parameters and the resulting target parameters of the coating a model, called "two level factor analysis", constructed on mathematical statistics /2-6/ can be applied to model the microstructure of the coating in the plasma spray process. The method was previously applied with sucess in plasma welding processes /7/ and in plasma spraying of metallic and ceramic coatings /8-11/. In the work presented here a brief introduction into the principles of two-level factor analysis is given. The method is then applied in optimising the atmospheric pressure plasma-spray process (APS-process) for newly developed τ-boride Ni-hardfacing alloys of the type Ni-Cr-(R_m)-B, where R_m represents the refractory metal Nb, Ta, or Ti. Plasmasprayed coatings made of powders of the new nickelbase hardfacing alloys can be used extensively to protect exposed surfaces of technical components against wear in corrosive environments. These alloys possess a good abrasion resistance, due to their hard phases, especially their complex-boride phases (eg. τ-phase = $(Ni,R_m)_{23}B_6$). The hardphases are embedded in a ductile nickel solid solution matrix. Due to the chromium content a good resistance against alkaline solutions, nitric and sulphuric acids, and against oxidation at higher temperatures /12-15/ is achieved. The main field of application for the highly boridic Ni-Cr-Ti-B coatings is wear protection while for the less boridic Ni-Cr-Nb-B and Ni-Cr-Ta-B coatings it is a combined wear and corrosion protection. The coating thickness chosen was between 400 and 600 μm.

Materials and Methods

Materials:
The composition, melting characteristic and densities of the new hardfacing powders used to manufacture coatings on substrates of mild steel are listed in Table 1.

Table 1: Chemical compositions, liquidus temperatures, melting intervals and densities of the coating materials used

Material	Composition Ni	Cr (wt.%)	Ta	Nb	Ti	B	C	Melting-Charact. T_{liq} (°C)	ΔT	Density ρ (g/cm³)
Ni-Cr-Nb-B										
75/05	72.1	9.4	-	15.6	-	2.9	.020	1115	57	8.21
75/K9	70.1	19.4	-	8.8	-	1.7	.022	1131	41	7.91
75/K10	66.8	19.3	-	10.6	-	3.3	.04	1152	73	7.95
Ni-Cr-Ta-B										
76/01	56.9	7.3	34.8	-	-	1.0	.015	1125	43	10.22
76/K2	65.9	8.3	24.1	-	-	1.7	.014	1149	37	9.60
76/K4	65.3	9.8	21.2	-	-	3.6	.11	1144	38	8.80
Ni-Cr-Ti-B										
121	65.2	11.3	-	-	18.6	4.9	-	1227	127	7.28
124	74.3	11.8	-	-	8.9	5.0	-	1179	73	7.59
127	78.3	11.3	-	-	4.7	5.7	-	1184	153	7.53
129	68.3	13.9	-	-	11.7	6.1	-	1148	51	7.40
214	62.9	21.3	-	-	12.0	3.8	-	1175	70	7.59
220	57.5	21.3	-	-	16.1	5.1	-	1218	164	7.30
227	64.4	23.3	-	-	5.4	6.9	-	1168	82	7.43
229	56.5	24.8	-	-	12.2	6.4	-	1170	109	7.20

The pre-alloyed powders were produced in an atomizing process in argon by the company H.C.Starck, Berlin in the plant at Laufenburg (Germany). Due to the process the powder grains are of an almost ideal spherical shape and possess good flow properties. The grain-size chosen for the experiments was between 22.5 µm and 45 µm.

The oxygen content measured in the powders ranged from 300 to 1400 ppm. The three types of powders show liquidus temperatures between ≈1100 and ≈1200°C and chromium contents of ≈10 or ≈20wt.%.

The boron content of the less boron containing Ni-Cr-Nb-B and Ni-Cr-Ta-B alloys varies between 1.0 and 3.6 wt.%, the niobium or tantalum content between 8.8 and 34.8 wt.%, and the densities are about 8 g/cm^3 for the first and 9-10 g/cm^3 for the second alloy system. These powders, designed for wear resistance in corrosive environments, show a low number of evenly distributed, finely precipitated, supersaturated hardphases, such as Ni_xB, τ-boride and $Ni_x(R_m)_y$ phases, embedded in a low-boron containing ductile nickel solid solution matrix. Typical powder-microstructures are shown in fig. 1a-c.

The high boron containing Ni-Cr-Ti-B powders have boron contents between 3.8 and 6.9 wt.%, titanium contents between 4.7 and 18.6 wt.%, and chromium contents of either ≈10 or ≈20 wt.%. The density range of these alloys (between 7.2 and 7.6 g/cm^3) is lower than for the previous powders. Typical microstructures of the powders are shown in fig. 1d. The high-boron-containing Ni-Cr-Ti-B particles possess microstructural features desired in wear protection, such as evenly distributed precipitations of TiB_2, chromium boride hardphases, and a high content of the τ-boride phase. The large amount of hardphases are embedded in a ductile matrix. The main differences in the microstructural features of the Ni-Cr-Ti-B powder compositions are the varying ratio of hardphases to matrix and the chromium contents.

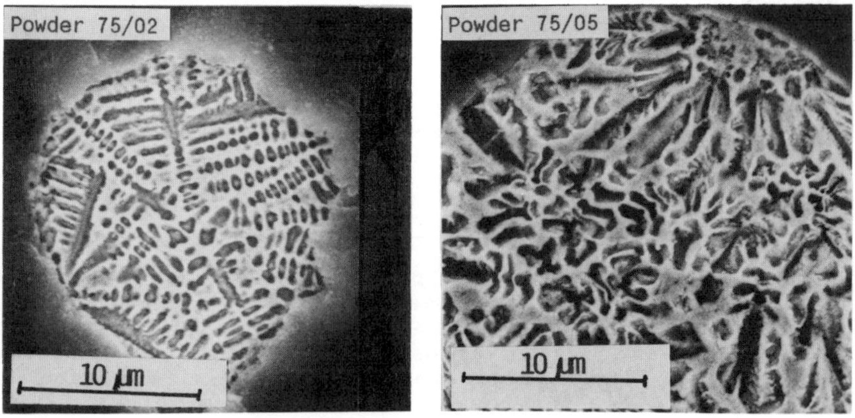

(a) Ni-Cr-Nb-B (75/05, 75/K9, 75/K10, 75/02). Supersaturated phases of: Ni-(solid solution), Ni_3B, (Cr_2B, τ, Ni_3Nb)

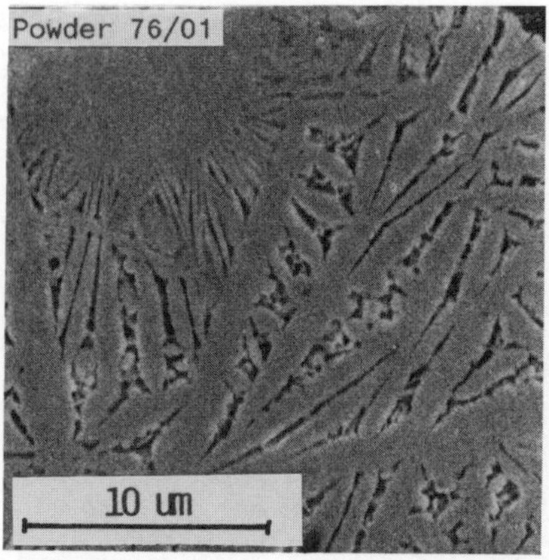

(b) Ni-Cr-Ta-B (76/01). Supersaturated phases of: Ni_3Ta, (Cr_2B, τ)

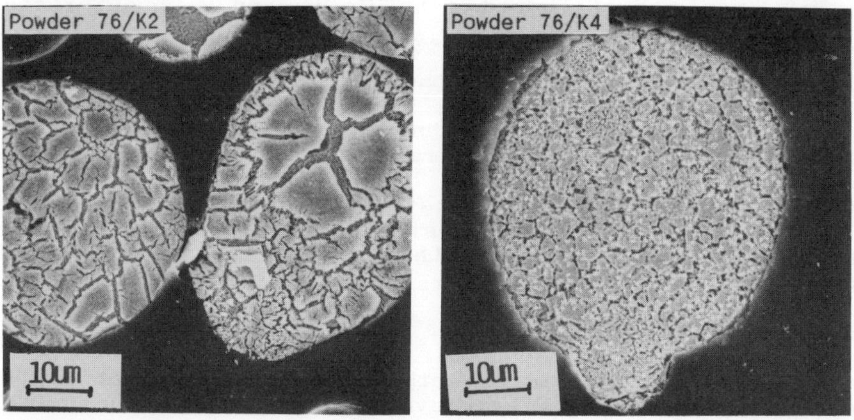

(c) Ni-Cr-Ta-B (76/K2, 76/03, 76/K4). Supersaturated phases of: Cr_xB_y (τ, Ni-(solid solution), Ni_3Ta)

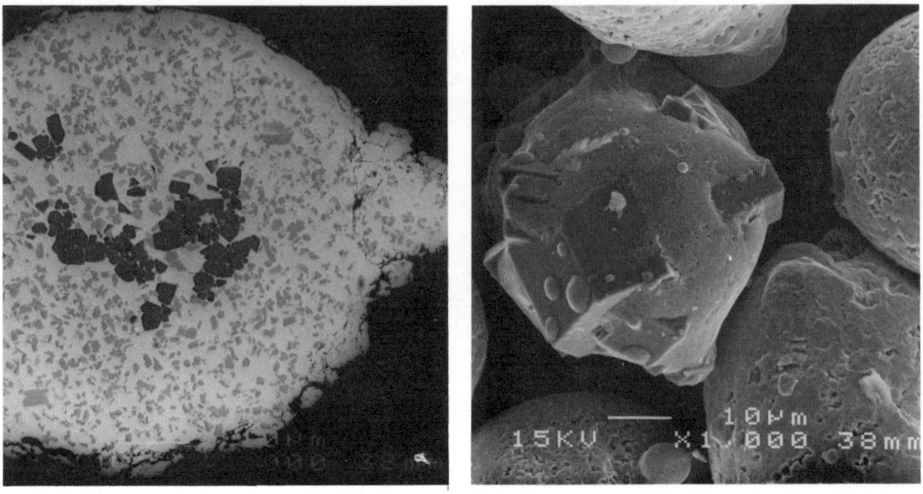

(d) Ni-Cr-Ti-B. TiB_2 (black cubes), supersaturated phases of evenly distributed fine Cr_xB_y and τ presipitates, and Ni-(solid solution)

Fig. 1a-d: Typical shape and microstructures of the powders

Principles and application of the two-level factor analysis: The experimental parameters required to produce the desired coating microstructure have been determined by applying a statistical model. The first stage is to set up a test-plan (termed "two-level factor analysis") on the basis of known zero point X_0 for the individual experimental parameters as determined in pre-experiments.

In two-level factor analysis the individual experimental parameters, X_i, are varied around a fixed zero point X_{01} with a constant variation interval ΔX_i. By suitable standardisation on x_i the two levels receive the values -1 and +1, which simplifies the statistical evaluation of the test series (factor plans). The total number of tests to be carried out is $dE=2^k$, where k is the number of parameters x_i. The plan matrix for a factor plan with three variables is shown in table 2.

Table 2: Standard plan matrix for a factor plan with 8 tests, three standardized influential variables (x) and their interactions ij ($x_2=x_1x_2$, $x_3=x_1x_3$, $x_3=x_2x_3$), and ijk ($x_3=x_1x_2x_3$) to study the resulting target quantities (Y)

test	plan matrix			matrix of independent variables X_i								matrix for target quantities (Y)
	x_1	x_2	x_3	0	1	2	interaction 12	3	interaction 13	interaction 23	interaction 123	y_1 y_2 y_3
1	–	–	–	+	–	–	+	–	+	+	–	
2	+	–	–	+	+	–	–	–	–	+	+	
3	–	+	–	+	–	+	–	–	+	–	+	
4	+	+	–	+	+	+	+	–	–	–	–	
5	–	–	+	+	–	–	+	+	–	–	+	
6	+	–	+	+	+	–	–	+	+	–	–	
7	–	+	+	+	–	+	–	+	–	+	–	
8	+	+	+	+	+	+	+	+	+	+	+	
effect matrix – regression coefficients: b_i, b_{ij}, b_{ijk}												

The standard plan matrix shows clearly how not only the principle factors but also two-factor (ij) and three factor interactions (ijk) affect the target quantity. Interactions of higher order fall often in the range of test scatter and thus have no effect on the target quantity. This simplifies the process. The target quantities (Y) are to be measured after the tests performed.

If the influence of more than four factors is to be investigated, the number of necessary tests increases considerably. The relatively high number of tests demanded in such cases can be reduced by mixing principal and interaction effects of higher order /10/. Mixed test plans are generally called 2^{k-p}-part factor plans. Here k is the total number and p the number of principal factors newly introduced into the plan. Table 3 shows the mixing of a 2^3-factor plan with an extra two, three or four factors.

Table 3: Mixed effects in various part factor plans

	matrix of the independent variables							
2^{5-2}	0	1	2	5	3			4
	Interactions	25	15	12	45	13	14	35
				34		24	23	
2^{6-3}	0	1	2	5	3	6		4
	Interactions	25	15	12	16	13	14	23
		36	46	34	45	24	56	35
2^{7-4}	0	1	2	5	3	6	7	4
	Interactions	25	15	12	16	13	14	17
		36	37	34	27	24	23	26
		47	46	67	45	57	56	35

In addition to the tests covered by the factor plan, additional tests are carried out in the zero points of the factors (X_{01}) and the resulting target quantities (Y) are compared to the original central point value y_0 of the target quantities. These tests are spread over the entire test period to detect possible changes in the system, and to verify the reproducibility of the test procedure. Furthermore, with the central point value y_0 (a target quantity) obtained it is possible to determine whether the resultant regression model (see eq. 1, below) reflects the real curve with sufficient accuracy. The test plan is carried out at least twice, in order to be able to evaluate the measurements statistically.

The objective of factor analysis is the creation of a functional relationship between the target quantity y and the variables x_i. This describes the regression model (1)

$$\begin{aligned} y&=b_0+b_1x_1+b_2x_2+b_{12}x_{12}+\dots \\ &=b_0+db_ix_i+db_{ij}x_{ij1}+\dots \end{aligned} \tag{1}$$

In this polynomial all principal (b_i) and two-factor-interaction effects (b_{ij}) are included. In test plans with more than four factors we limit ourselves only to the significant effects. The coefficients b_i and b_{ij} are determined from the plan matrix (see table 2) and tested for their significance by mathematical statistical methods. The calculations are in detail explained in /10/.

In most cases it is observed that the relation $y=f(x_1,x_2,\dots x_i)$ represents a curved surface. Each 2^{k-p}-

part factor plan occupies only 2^{k-p}-points on this surface, among which the desired optimum is not usually present. The model equation makes it possible to calculate lines of equal y-values and to set up a vector perpendicular to these, which points in the direction of the extreme value. Further tests are carried out along this vector, until the response increases no further. This point represents the central point for a further vector plan, within the variation range of which the optimum must lie.

Further detailed descriptions of the method and the formulae and procedures necessary for the evaluation can be found in /1-6,10/.

Application of two-level factor analysis to plasmaspraying of new hardfacings

Two level factor analysis has found practical application in the investigation of the influences of spraying process parameters on the quality of 400-600µm thick APS coatings of newly developed Ni-hardfacing alloys. These alloy-powders are designed for wear protection, especially for those applications where resistance against corrosion from aqueous acidic or alkaline solutions or/and oxidation resistance at higher temperatures is also desired.

In the spraying process it is necessary to produce dense coatings (low porosity) with a defined thickness and to retain the microstructure of the powders in terms of their compositions and hardphases formed, where possible. Melting of the hardphases, especially the chromium-boride hardphases, during the coating process has to be avoided. These phases do not precipitate naturally in the APS-process due to the high cooling rates governing the process. A reduced amount of precipitated hard phases leads to a decreased wear resistance and increased embrittlement of the coating resulting in crack-formation perpendicular to the surface. This latter phenomenon occurs when a high amount of boron is dissolved in the matrix-solid solution.

In order to obtain a good wear resistance the high boron containing Ni-Cr-Ti-B coatings should further contain fine, evenly distributed, precipitated boridic particles, mainly of the type Cr_xB_y, TiB_2, and $(Ni,Ti)_{23}B_6$. The particles should be embedded in a ductile low-boron-containing nickel-chromium-solid solution matrix. To obtain a combined corrosion and wear resistance the lower boron containing Ni-Cr-Nb-B/ Ni-Cr-Ta-B coatings should not only be dense, but the boron containing hardphases precipitated should be of the type $Ni_x(R_m)_y$, Ni_xB_y, and $(Ni,R_m)_{23}B_6$ (but not Cr_xB_y), in order to maintain a high chromium and low boron content in the nickel-chromium matrix.

A further microstructural aspect of these APS coatings is their oxide film content. The oxide films have a severely negative influence not only on the ductility and the thermal expansion behaviour of the coating, but on the adhesion of the coating to the substrate and the resulting bondstrength. Furthermore they reduce the corrosion resistance of the coating, due to the very different electronegativity of the metallic, partially oxidized and the oxidized phases. During the spraying process oxide film formation has therefore to be kept to a low level.

Quantifiable coating properties:
Easily measurable and quantifiable coating properties are primarily the coating thickness, and the porosity, and secondarily the oxide-film content, the surface roughness, the degree of powder-melting, the hardness, etc. This examination concentrates mainly on the primary target quantities "thickness" and "porosity".

Selected experimental parameters:
For the first investigation period a mixed 2^{6-2}-part factor plan was devised, in which six selected experimental parameters were varied to study their influences on thickness and porosity of the coatings. The parameters selected were the current intensity (x_1), the powder mass-stream (x_2), the argon carrier gas flow rate (x_3), the spraying distance (x_4), the torch advance speed (x_5), and the number of passes (x_6), see table 4. The values for further experimental parameters as nozzle geometry, plasma-gas composition and flow rate were chosen from the pre-experiments.

Table 4: Selected experimental parameters and range

Parameter	variation range	
	Ni-Cr-Nb-B/Ni-Cr-Ta-B	Ni-Cr-Ti-B
power:	25-37 kW	45-50kW
nozzle diameter, mm:	8	6
plasma-gases (90SLPM):	25-50Ar,3-36He,0-3H_2	30-70Ar,10-20He,0-10H_2
x_1 current intensity:	500-600 A	450-650
x_2 powder mass-stream:	8-17 g/min	6-12 g/min
x_3 powder feeding-gas:	4-6 SLPM argon	4-6 SLPM argon
x_4 spraying distance:	100-140 mm	100-120 mm
x_5 torch advance speed:	0.8-1.5 m/min	0.6-0.9 m/min
x_6 number of passes:	4-7	2-7

Calculation of coating thickness and porosity:
Using statistical test planning the significant and non-significant influences on the test quantities "porosity" and "coating thickness" are determined. From this regression models for the significant quantities can be created, with which coating thickness and porosity can be theoretically calculated. The model equations must be subjected to an adequacy test, in order to ensure that the model values correspond to reality. These model equations of course only have validity in the test area investigated. Within this test area, therefore, any quantities of coating thickness and porosity can be determined and the parameter values necessary for these mathematically determined. However, when using any regression model, care must be taken to ensure that the test and the subsequent production conditions are identical. In the following the regression equations are listed for the 2^{6-2}-part factor plan (eq. 2 and 3) and the 2^3-factor plan (eq. 4 and 5) in the optimum. The indices S and P indicate coating thickness and porosity:

a) 2^{6-2}part-factor plan

$$y_S = b_0+b_2x_2+b_3x_3+b_4x_4+b_5x_5+b_6x_6+b_{13}x_{13}+b_{16}x_{16}+b_{24}x_{24}, \quad (2)$$
$$y_P = b_0+b_1x_1+b_2x_2+b_5x_5+b_{12}x_{12}+b_{13}x_{13}+b_{15}x_{15}+b_{16}x_{16}, \quad (3)$$

b) 2^3 -factor plan

$$y_S=b_0+b_2x_2+b_3x_3 \quad (4)$$
$$y_P=b_0+b_2x_2+b_{13}x_{13}+b_{23}x_{23}+b_{123}x_{123}. \quad (5)$$

The coefficients b_j are the significant regression coefficients, whilst b_0 is the mean of the test results.

Results and Discussion

Porosity:
The APS-process produces coatings with a large variety of complex microstructural features. Fig. 2a,b shows examples of coating microstructures obtained with the same powder but different process parameters in Ni-Cr-Ti-B coatings sprayed with low porosity. Pre-experiments, performed to establish a starting parameter set for two-level factor analysis, showed two possible ways to obtain a dense coating. The one is by complete melting of the powder particles by choosing a low powder mass stream and a low torch advance speed (fig. 2a). With this parameters the resulting microstructure is dense, but the complete melting of the powder particles leads easily both to an undesirable dissolution of the chromiumboride hardphases, leaving only the larger and thermally more stable TiB_2-phase, and a pronounced oxide-film formation. By choosing a higher cur-

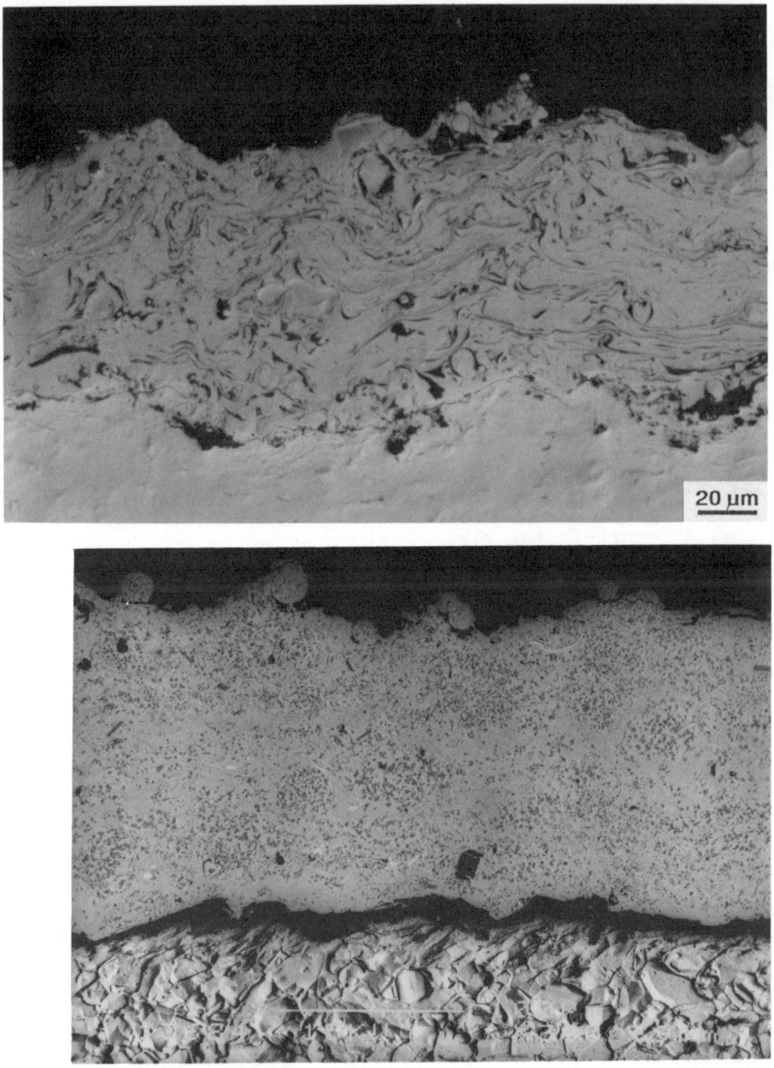

(a)

(b)

Fig. 2a,b: Complexity of microstructural features of dense sprayed Ni-Cr-Ti-B coatings. (a) Dense coating, complete melting of powder particles, dissolved chromium-boride hardphases, oxide-films. (b) Dense coating with desirable chromium-boride and τ-boride hardphases evenly distributed in the nickel-rich matrix and a low level of oxide films rent intensity of 650 A instead of 450 A, and using a higher plasma velocity by reducing the nozzle diameter from 8 mm to 6 mm, changing the plasma gas composition from 30Ar, $4H_2$, 20He to 70Ar, $10H_2$, 10He and increasing the plasma-gas stream to 90 l/min a dense coating with the desirable chromiumboride and τ-boride hardphases evenly distributed in the nickel-rich matrix and a very low level of oxide films, is achievable as shown in fig. 2b.

Changing the parameters meant the particles were exposed to the hot zone of the plasma and to the atmosphere for shorter periods. However, to ensure the powder surfaces and matrix were melted sufficiently the burner power was increased from 40 to 45-50 kW.

Higher plasma temperatures achieved by increasing the helium content (to 36 parts) of the plasma gas resulted in a higher degree of powder melting for the Ni-Cr-Nb-B and Ni-Cr-Ta-B coatings examined, a lower porosity of the coatings, and simultaneously retaining the desired microstructure of the coating, as shown graphically in Fig. 3. The correlation of degree of powder melting with porosity and surface roughness are shown for Ni-Cr-Nb-B and Ni-Cr-Ta-B coatings in Fig. 4a and 4b, respectively.

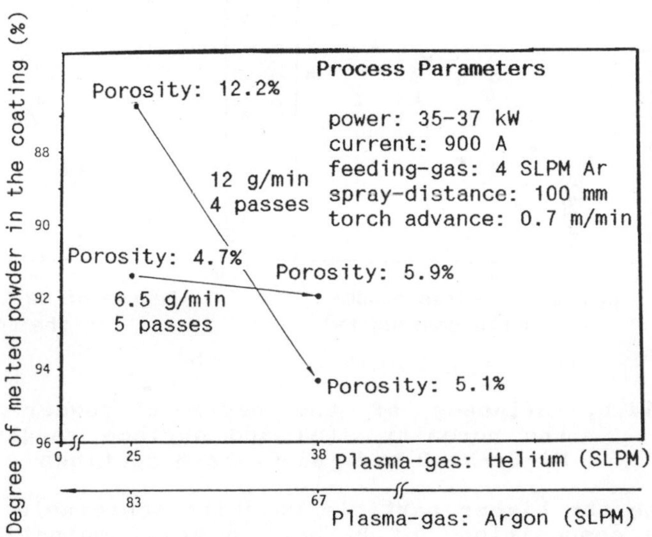

Fig. 3: Influence of plasma-gas composition on the degree of powder melting and porosity of Ni-Cr-Nb-B and Ni-Cr-Ta-B coatings

The investigations showed, that a good coating quality (i.e. low amount of pores primarily and often combined with it the secondary target quantities such as few oxide films, the desired coating microstructure, and low surface roughness) is mainly governed by the process-parameters such as the atmosphere/plasmagas composition in which the coating process takes places and the particle velocity achieved in the process.

Furthermore the more inert the atmosphere, and the higher the helium content (in the argon/helium plasma gas), the lower is the amount of undesirable impurities in the coating. The higher the particle velocity, the better is the adhesion, the lower is the coating porosity, the better is the preservation of hardphases, and the lower is the oxide content. The main factors governing the particle velocity are the shape of the nozzle and the combination of plasma-gases used.

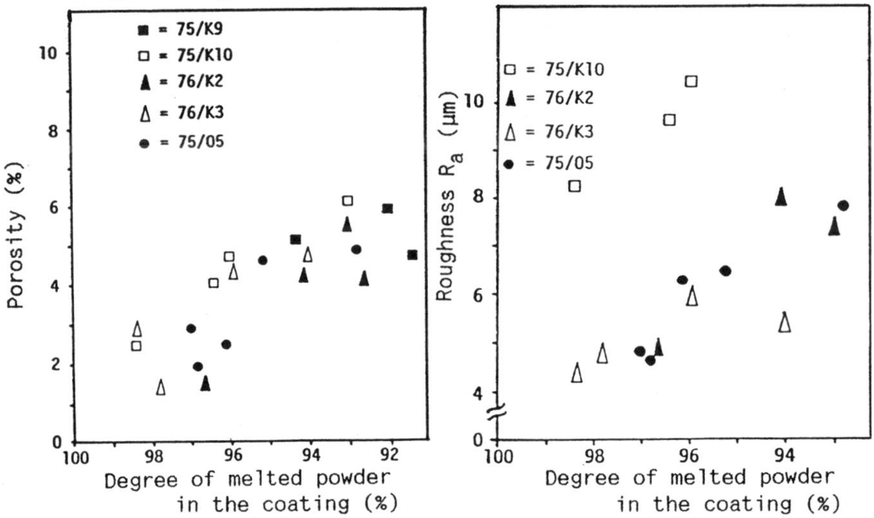

(a) (b)

Fig. 4a,b: Influence of the degree of powder melting on the porosity (a) and surface roughness (b) of Ni-Cr-Nb-B and Ni-Cr-Ta-B coatings

Despite the higher particle velocity achievable in a plasma-gas composition of H_2/He, a gas combination of Ar/He proved to be more suitable. The H_2 content leads to adsorption of hydrogen on the surface of the coating particles and then to an absorption of hydrogen into the nickel solid solution matrix, resulting in turn in hydrogen embrittlement of the coating.

Coating thickness:
Effects as adhesion, internal stresses, and porosity are amongst other effects directly or indirectly governed by the coating thickness per pass. The main influence on the coating thickness comes from the parameters powder mass stream (Fig. 5), powder feeding gas and the number of passes. These parameters have a positive influence, i.e. the coating thickness increased with increasing value of the parameters. The parameters spraying distance and torch advance speed have a negative influence. The variable current intensity does not have a direct effect on the coating thickness.

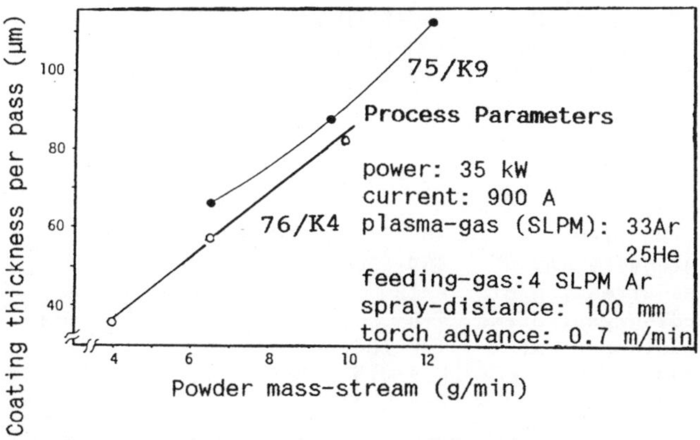

Fig. 5: Influence of the process-parameter powder mass stream on the coating thickness

The preliminary test results were drawn into the 2^{6-2}-part factor plan in order to choose from the multitude of spray-parameters those which may have the greatest influence on the coating. The plan was constructed in such a way that despite the mixings the principal effects of the two factor interactions could be determined separately. In addition to the tests involved in the factor plan, zero point tests also took place, on two occations. Table 5 shows the selected parameters, their variation range and the influences of the parameters on coating thickness and porosity. By regression analysis and significance testing the variables with most influence on the spraying process were determined.

Table 5: Influence of the parameters on the coating thickness and porosity in the operating zone of the 2^{6-2}part factor plan

parameter	variation range	influence on thickness	influence on porosity
Ni-Cr-Nb-B/Ni-Cr-Ta-B:			
power, kW :	25-37		
nozzle diameter, mm:	8		
plasma-gases, 90 SLPM :	25-50Ar,3-36He,0-3H_2		
x_1 current intensity, A:	500-600	0	0(-)
x_2 powder mass-stream, g/min:	8-17	+	0(+)
x_3 powder feeding-gas, SLPM:	4-6 SLPM argon	+	0(+)
x_4 spraying distance, mm:	100-140	-	0
x_5 torch advance, m/min:	0.8-1.5 n	-	+
x_6 number of passes:	2-7	+	0(+)
Ni-Cr-Ti-B			
power, kW :	45-50		
nozzle diameter, mm:	6		
plasma-gases, 90 SLPM :	30-70Ar,10-20He,0-10H_2		
x_1 current intensity, A:	450-650	0	-
x_2 powder mass-stream, g/min:	6-12	+	0 (+)
x_3 powder feeding-gas, SLPM:	4-6	0(+)	0
x_4 spraying distance, mm:	100-120	-	0(+)
x_5 torch advance, m/min:	0.6-0.9	-	+
x_6 number of passes:	2-7	+	0

Starting from the results of the 2^{6-2}-part factor plan, stepwise tests took place, which led in the direction of the optimum for coating thickness and porosity. Because the regression equations point in different directions, separate stepwise tests were necessary for each of the target quantities. These tests show that for the low-boron-containing Ni-Cr-Nb-B and the Ni-Cr-Ta-B powders the number of passes and the torch advance speed must both be small, in order to achieve coating thicknesses within the required range and to minimise the inherent stress in the coatings. It is necessary to adjust the current intensity to the range around 600 A and to keep the powder mass stream high.

The target quantity "porosity" is combined with secondary target quantities, as shown, and therefore the most important variable governing the wear and corrosion resistance of the coating. Thus the new parameter range for the factor plan to establish the optimum conditions started from the results of the stepwise tests for porosity. The parameters chosen here were the significant variables of the 2^{6-2}-factor plan (current intensity, powder mass-stream and torch advance speed). With these three values a 2^3-factor plan was created. The variation ranges of the parameters and their influence on the coating quality are shown in Table 6. The best values as determined in pre-experiments and in the 2^{6-2}-part factor-plan were used for the other process parameters (nozzle, powder feeding gas, spraying distance, plasma-gases).

The optimum values of the parameters describing the coating quality are a thickness per pass of 140 µm, and a porosity of 1-3%. The resulting surface roughness R_a was 4-6µm, the degree of powder-melting was 98%, and the hardness of the coatings measured was 750±150 $HV_{0.2}$.

Table 6: Parameters, variation ranges and influences of the parameters on the coating thickness and porosity in the test range of the 2^3 part factor plan to determine the optimum values

parameter	variation range	influence on thickness	influence on porosity
Ni-Cr-Nb-B (75/05)			
Ni-Cr-Ta-B (76/01)			
x_1 current intensity:	600±50 A	0	0
x_2 powder mass-stream:	15±2 g/min	+	0(+)
x_3 torch advance speed:	1.0±0.2 m/min	-	0(+)

A comparison of the principle effects in the 2^3-factor plan around the optimum with those in the 2^{6-2}-part plan showed that the effects of the parameter variation on the measured microstructure parameters describing the coating quality in the range of the 2^3-factor plan are all so small that it can be assumed that the parameter values for the minimum achievable porosity lie in the test area.

A summary of the measured coating properties obtained with the new Ni-hardfacing alloys after optimisation of the spray process-parameters is given in table 7. The corresponding microstructures and morphologies of the plasmasprayed wear and corrosion resistant Ni-Cr-Nb-B, Ni-Cr-Ta-B, and Ni-Cr-Ti-B coatings are shown in Fig. 6a-c.

(a) Ni-Cr-Nb-B coating (75/05)

(b) Ni-Cr-Ta-B coating (76/01)

c) Ni-Cr-Ti coating (227).
Etched in 4 parts HNO_3, 1 part CH_3COOH, 1.5 parts H_2O

Fig. 6a-c: Microstructures and morphologies of plasmasprayed wear and corrosion resistant Ni-Cr-Nb-B, Ni-Cr-Ta-B and Ni-Cr-Ti coatings

Table 7: Coating properties after optimisation

coating properties	Ni-Cr-Nb-B/ Ni-Cr-Ta-B	Ni-Cr-Ti-B
coating thickness per transition, µm	120-140	150-250
porosity,%	1-3	0.5-1
surface roughness R_a(µm)	4-6	4-6
hardness ($HV_{0.2}$)	600-900	900-1300

Conclusion

The use of thermal sprayed coatings as protection against wear in corrosive environments requires a precise knowledge of the influence of the spraying process parameters on the coating properties (primarily thickness and porosity, and secondarily adherence, hardness, surface roughness, microstructure, oxide film content etc.) to ensure the required wear and corrosion resistance.

Of great interest in optimizing the process parameters are not only the principal effects of one variable parameter, but also the interactions between two or three influential variables. Two-level factor analysis offers an elegant method of determining these interactions, but also the principal influences. With the regression equations obtained, within the test area the coating properties selected as target quantities can be calculated in advance for any parameter values or vice versa, and those process parameters adjustments determined, which lead to the desired coating properties. The two-level factor analysis is a tool to analyse the plasma spraying process with minimum effort and adequate accuracy. Detailed statements can be made with two-level factor analysis of higher order or with the three-level factor analysis. However, the test effort increases considerably in these cases.

Acknowledgement

Special thanks are directed to the German Federal Ministery for Research and Technology (BMFT Nr. 13 N 5467 9 / BMFT 03T00221) and the company H.C.Starck-Berlin for financal support of this research.

References

1. I.A. Fischer, "Variables influencing the characteristics of plasma-sprayed coatings", *International Metallurgical Review*, 1972, Vol. 77.
2. E. Scheffler, *Einführung in die Praxis der statistischen Versuchsplanung*, VEB Deutscher Verlag für Grundstoffindustrie, Leipzig 1986.
3. H. Bandemar, A. Bellmann, *Statistische Versuchsplanung*, Reihe MINÖVA, Bd. 1912, Verlag Harry Deutsch, Thun und Frankfurt/Main, 1979.
4. G.E.P. Box, W.G. Hunter, J.S. Hunter, *Statistics for Experimenters*, John Wiley & Sons Inc., New York-London-Sydney, 1978.
5. J.N. Bornstein, K.A. Semendjajew, *Taschenbuch der Mathematik*, Verlag Harry Deutsch, Thun und Frankfurt, 1981.
6. J.W. Linnik, *Die Methode der kleinsten Quadrate in moderner Darstellung*, VEB Deutscher Verlag der Wissenschaften, Berlin 1961.
7. A. Ait-Mekideche, Dissertation, RWTH Aachen, 1989.
8. P. Lu, Dissertation, RWTH Aachen, 1989.
9. R. Dekumbis, P. Huber, M. Villat, *MCrAlY vacuum plasma-sprayed coatings: factorial two-level experiments as an aid to determine the influence of spraying parameters on coating properties*, DVS-Berichte Bd. 80, Düsseldorf 1980, 153-160.
10. D. Hofmann, *Laser-Veredelung von Plasma-Spritzschichten*, VDI-Forschritttsbericht Nr. 240, Reihe 5: Grund- und Werkstoffe, VDI-Verlag, Düsseldorf, 1991.
11. E. Lugscheider, D. Hofmann, D. Schubert, A.R. Nicoll, *Application of two-level factor analysis to the optimisation of thermal spray coatings using the example of plasma spraying*, WELDING AND CUTTING 10, 1991, E217-219.
12. E. Lugscheider, B.C. Oberländer, *A comparison of the properties of coatings produced by laser cladding and conventional methods*, to be published in Materials Science and Technology, 1992.
13. B.C. Oberländer, *Oberflächenveredelung mit tantal- bzw. niobhaltigen Nickelhartlegierungen am Laserstrahl*, Reihe Schweißtechnische Forschungsberichte Bd. 45, DVS-Verlag, Düsseldorf, März 1992.
14. A. Sabour Rouhaghdam, Dissertation, RWTH Aachen, 1992.
15. E. Lugscheider, A. Krautwald et al., *Surface and Coating Technology*, 32, 1987 273-284.

Chapter 6

PROPERTIES AND APPLICATIONS OF PLASMA SPRAYED CERAMIC COATINGS

CHUANXIAN DING, ZHAOHE TONG AND DONGSHENG YAN
Shanghai Institute of Ceramics, Chinese Academy of Sciences
1295 Dingxi Road, Shanghai 200050, P. R. China

ABSTRACT

Plasma spaying has received considerable attention as a process for obtaining protective coatings. In this paper plasma sprayed ceramic coatings such as thermal barrier, wear resistance, infrared radiation and biological coating materials as well as diamond films are presented. The physical, mechanical, thermal and some other properties of the coatings are reviewed. The microstructural features of the coatings are also described. Examples of applications of the ceramic coatings in various industrial sectors are illustrated.

1. Introduction

With the continuing development of industrial manufacturing sectors, materials are being subjected to severe and more demanding environments. Plasma spraying of ceramic materials becomes a subject of intense interest and affordable solution for many industrial problems and applications (wear resistance, thermal barrier, infrared radiation, biological compatibility etc.). As the fundamentals of plasma spraying are easy to be found in the textbook and the development trends in technology can be found elsewhere [1]. This paper will deal with the ceramic coatings frequently used in industries. The physical, mechanical, thermal and some other properties of ceramic coatings are reviewed. The microstructural feature of the coatings is addressed. The relationships between the microstructure of the coatings and their properties are discussed. Table 1 summarizes the ceramic coatings that have been developed and widely used in various branches of industry.

2. Characteristics and Properties of the Ceramic Coating

2.1. Microstructure of Ceramic Coatings

The microstructure of a plasma sprayed coating in general is lamellar in nature

Table 1. The main application features of ceramic coatings

Coatings	Engineering environment	Engineering reference application
Al_2O_3 Al_2O_3-TiO_2 Cr_2O_3 WC-Co Cr_3C_2- NiCr	wear	mechanical parts in textile industry, petrochemical and paper making equipments [2,3]
TiO_2-ZrO_2- Nb_2O_5	infrared radiation	electric heating [4]
ZrO_2 bioglass	biological compatibility	bioceramic coatings for artificial bones, joints and tooth roots [5]
ZrO_2 Al_2O_3 ZrO_2-metal	low thermal diffusivity good thermal shock resistance	thermal barrier coatings for diesel engine and aircraft parts[6]
NiCr NiAl NiCrAlY Mo	high strength good adhesion to substrate	bonding coating between the ceramic coating and substrate [7]

formed by the flattened platelet of individual particles. Fig. 1 illustrates a schematic diagram of the lamellar structure. There are defects, such as microcracks and different kind of pores, in the ceramic coatings, which affect and are closely associated with the properties of the coatings and their applications.

Owing to very rapid quenching and solidification during coating formation, sometimes, there are different crystallographic modifications found in such ceramic coatings. For instance, some metastable high temperature crystallographic

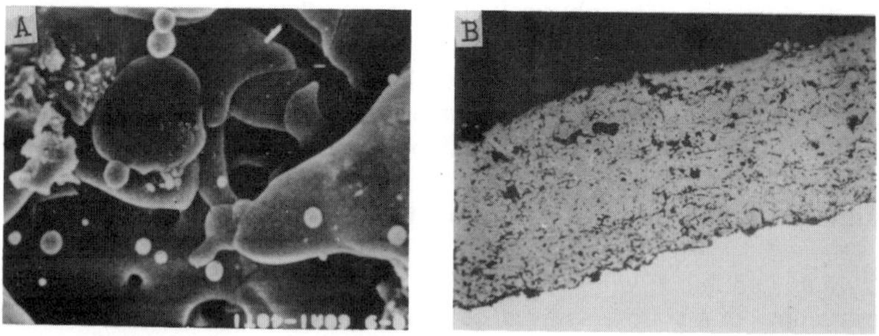

***Figure 1** Typical structure of ceramic coatings: (A) surface of an Al_2O_3-TiO_2 coating and (B) cross section of a Cr_2O_3 coating (200x)*

polymorph may be frozen in the coatings. Plasma sprayed Al_2O_3 is one of such cases where the crystallographic phase composition occurred seems to be related to the thickness of the coating. In a thin Al_2O_3 coating (0.1mm thickness), the crystalline phase is mainly composed of γ-Al_2O_3 with some η and α-Al_2O_3. The content of γ-Al_2O_3 is usually over 70%. Whereas in a thick coating, because of the slower cooling rate and the poor thermal conductivity of already deposited material, the γ-Al_2O_3 content is usually less than 20% and the coating is mainly composed of α-Al_2O_3. Generally, in the Al_2O_3-TiO_2 coating, x-ray diffraction shows the presence of α-Al_2O_3, γ-Al_2O_3, rutile, anatase and spinel (Al_2TiO_5). Sometimes, other unknown "x-phases" were also observed [8].

2.2. The Physical and Mechanical Properties of Coatings

The physical and mechanical properties of the ceramic coatings were usually measured by the following methods: porosity and bulk density by water immersion, bend strength of the coatings by the three-point or four-point bend test. Table 2 lists the physical and mechanical properties of a number of the coatings. Plasma sprayed

2. The physical and mechanical properties of coatings

Coating materials	Porosity (volume%)	Bulk density (g/cm^3)	Bend strength (MPa)
Al_2O_3(type-1)	3.0	2.58	49.0
Al_2O_3(type-2)	5.7	3.23	76.0
Cr_2O_3	4.7	4.52	81.3
ZrO_2(5wt%CaO)	8.7	5.52	49.9
ZrO_2(8wt%Y_2O_5)	9.7	5.40	67.6
80wt%Al_2O_3-20wt%TiO_2	6.0	3.48	89.2
74wt%Al_2O_3-20wt%ZrO_2-5wt%Nb_2O_5	4.9	4.03	30.4
88wt%WC-12wt%Co	10.8	13.87	112.7
75wt%Cr_3C_2-25wt%NiCr	10.8	6.16	274.4
80wt%Ni-20wt%Al	5.7	6.42	343.0
53wt%NiAl-47wt%ZrO_2	7.1	6.27	169.5
50wt%W-50wt%ZrO_2	12.2	9.32	66.6
50wt%Mo-50wt%ZrO_2	9.8	6.80	112.7

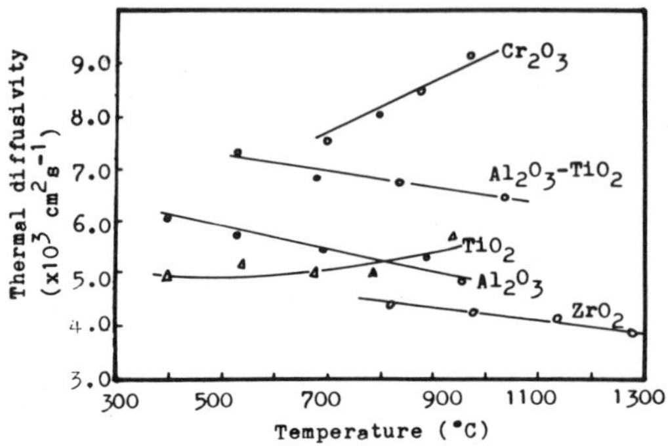

Figure 2 The thermal diffusivity of oxide coatings

ceramic coatings usually contain some porosity from 5 to 30% by volume. The thermal diffusivities of some ceramic coatings are shown in Fig. 2. The thermal diffusivity of most oxide coatings decrease with increasing temperature. However, the thermal diffusivities of Cr_2O_3 and TiO_2 coatings shows a reverse trend, increase with an increase of temperature. These characteristics are advantages with respect to the wear resistance especially when these coatings are used at elevated temperature. For example, the addition of TiO_2 to Al_2O_3 improves the thermal diffusivity of the coating and thereby improves wear resistance of such coating system. The thermal diffusivity of ZrO_2 coating is lower than that of other plasma sprayed ceramic coatings, and therefore makes it useful as a thermal barrier on metals.

Table 3. Thermal expansion coefficients of oxide coatings ($x10^{-6}\,°C^{-1}$)

Coating materials	Temperature(°C)				
	200	400	600	800	1000
ZrO_2-5wt%CaO	8.66	9.81	9.75	9.87	9.91
ZrO_2-18wt%Y_2O_3	7.33	8.76	9.34	10.00	10.53
Cr_2O_3	6.00	6.50	6.45	5.88	
80wt%Al_2O_3-20wt%TiO_2	5.57	6.59	6.47	5.56	
ZrO_2-NiAl graded coating	9.00	10.70	10.80	11.80	

From Table 2, it can be seen that a coating which contains a metal constituent usually possesses a higher strength. For instance, in the NiAl-ZrO_2 system, the bend strength is dependent on the content of NiAl. Besides, the addition titania to alumina results in considerable increase in bend strength of the coating. It may be that titania containing matrix in the alumina-titania coating contributes to the increase in the strength of the coating, The thermal expansion behaviour of coatings has also been investigated and the data are presented in Table 3.

3. Ceramic Coatings

Ceramic coatings can be classified according their applications into thermal barrier coatings, wear resistance coatings, infrared radiation coatings, biological coatings and diamond films. Some details about them are described as follows.

3.1. Thermal Barrier Coating

Plasma sprayed zirconia coating stabilized by CaO, MgO, Y_2O_3 or CeO_2 are usually used as thermal protection at high temperatures. The addition of a stabilizer agent can prevent the phase transition of ZrO_2 from tetragonal to monoclinic at elevated temperatures and thus improve the thermal shock resistance and corrosion resistance of ZrO_2 coatings [1,9]. Plasma sprayed ZrO_2 coatings have been used as thermal protection for aircraft engine parts such as combustor cans, nozzle and walls, gas path transition pieces, turbine blade platforms, after burner nozzle and flaps [10]. Recently, graded metal-oxide coatings with the ZrO_2 content being varied from 1 to 100% and thickness of 2 to 3 mm have been developed. Table 4 presents the results of thermal shock test for different kinds of coatings. The test results obtained indicated that the graded coating is more thermal shock resistant than the single and duplex coatings that were also tested. The reason for this improvement may be explained by the following two factors. The first is that the graded coating has a lower thermal expansion mismatch between the substrate and coating, and therefore

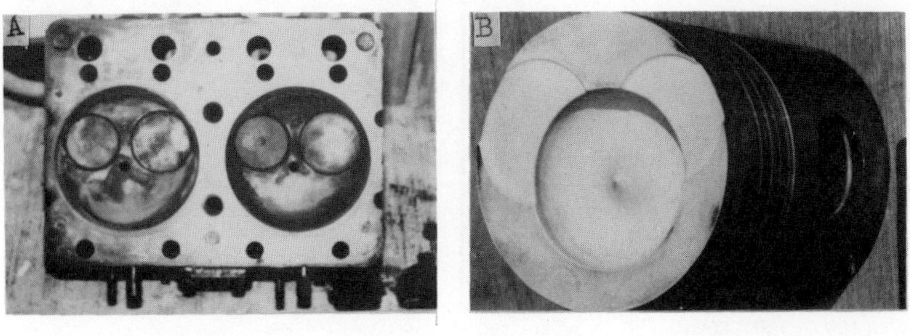

Figure 3 Insulated combustion chamber for diesel engine: (a) cylinder head and valves and (B) piston

internal thermal stress is reduced. The second point is that the metallic particles may play a role of impeding initiation and propagation of cracks and thereby improves the toughness of the coating. The graded coating can be used for thermal insulation of the combustion chamber of diesel engine. Fig. 3 shows the diesel engine parts coated with graded coating. Bench test for 200hr indicated that no spalling of the coating took place [11].

Expanded use of thermal barrier coating is being planned as parts of a space shuttle, commercial launch vehicle and space station. Aerospace applications are growing to include graded, plasma formed rocket nozzle and other high heat flux surface for radar absorbing, low observable coatings [10].

Table 4. Results of thermal shock test

Coating system	Average thermal cycles
pure stabilized ZrO_2 coating	8
Ni based alloy bond coating + stabilized ZrO_2 top coating	21
Ni based alloy +Stabilized ZrO_2 composition graded coating	30

3.2. Ceramic Coatings for Wear Resistance

Plasma sprayed ceramic and cermet coatings such as Cr_2O_3, Al_2O_3, TiO_2, Al_2O_3-TiO_2, WC-Co and Cr_3C_2-NiCr possess high surface hardness, low friction coefficient, good adhesive bonding to substrate and excellent diamond grindablity and are being developed to protect metal components. In order to understand more about their anti-wear properties, the wear and friction of some of these coatings have been Table
Table 5. The microhardness and friction coefficient of some coating materials

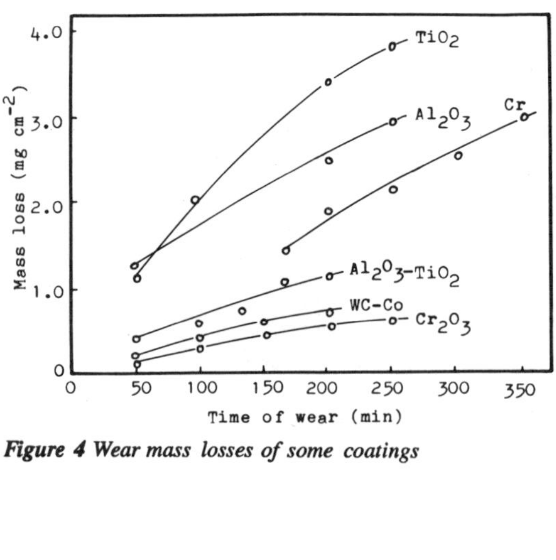

***Figure 4** Wear mass losses of some coatings*

Coating materials	Microhardness (GPa)	Friction coefficient (against stainless steel)
Al_2O_3-2	9.2	0.12-0.20
Cr_2O_3	10.0	0.14-0.15
TiO_2	8,2	0.10-0.15
80wt%Al_2O_3-20wt%TiO_2	10.4	0.10-0.11
80wt%WC-20wt%Co	12.1	0.11-0.13
75wt%Cr_3O_2-25wt%NiCr	10.5	0.13-0.15

investigated. The friction coefficients and microhardness values and wear mass losses are shown in Table 5 and Fig. 4 respectively. The wear mass loss curves indicated that the anti-wear capability of different coatings decrease in the following order: Cr_2O_3, WC-Co, Al_2O_3-TiO_2, Al_2O_3, Cr_3C_2-NiCr [12].

The stable wear rate of WC-Co and Cr_2O_3 coatings against steel were measured and plotted in Figs. 5 and 6. The following formula can be fitted for the WC-Co coating

$$W \alpha P^{0.61}$$

where W is the wear rate and P is the normal load, and for Cr_2O_3 coating

$$W \alpha P^{0.69}$$

The exponent values of normal load against wear rate for these two coatings are about the same and are near to 2/3, which is different from the value for bulk ceramic materials. This difference may be related to the different dominant fracture mechanism as a result of the lamella structure of the plasma sprayed coatings [13].

The worn surface and cross section of WC-Co and Cr_2O_3 coatings were detected using SEM. The results obtained were shown in Figs. 7 and 8, which showed that there were median cracks on the worn surface and cracks between layers in the coatings. The results obtained indicated that the median crack is the dominant factor with regard to the wear of ceramic

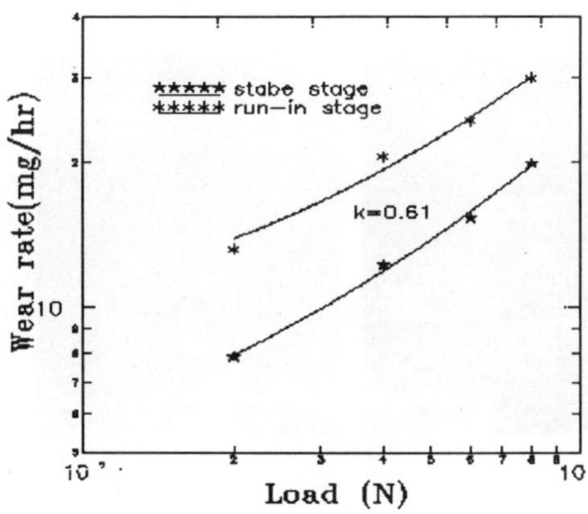

Figure 5 Logarithmic plot of wear rate of WC-Co coating vs. normal load

coatings. Because it has been found that the bending strength of the ceramic coating is 5-10 times greater than the cohesive strength between layers in the coating. Therefore, the cracks between layers in the coatings can be formed more easily than the median crack. The fractured particles are ready to leave from the contact surface when the median crack extends to the interface where the lateral cracks have already formed. Wear of the coatings happens when the fracture particles in the worn surface flake off. It has also been proved by Mendeson who investigated the wear of coating against grey cast iron [14].

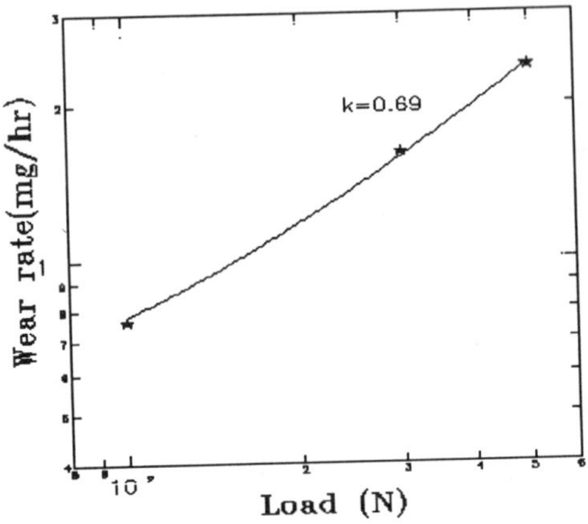

Figure 6 Logarithmic plot of wear rate of Cr_2O_3 coating vs. normal load

Plasma sprayed ceramic coatings for wear resistance become a subject of great interest and provide solution for many industrial problems, and have been used to resist abrasive, fretting, cavitation and erosion wear. Fig. 9 shows some mechanical parts such as sealing rings, guide rolls, pump sleeves and compressor cylinder in chemical, textile and mechanical industries. Potential advantages of wear resistant coatings made them be widely used to prolong the service life, reduce the cost of mechanical parts and protect the environment against pollution.

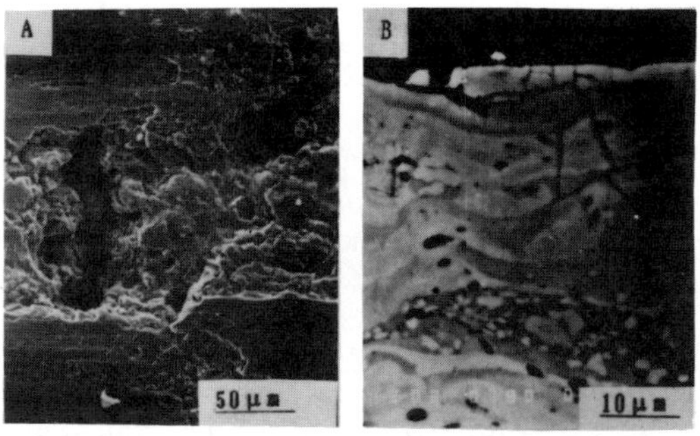

Figure 7 SEM micrographs of Cr_2O_3 coating: (A) flaking on the worn surface and (B) cracks and fracture from cross sectional view

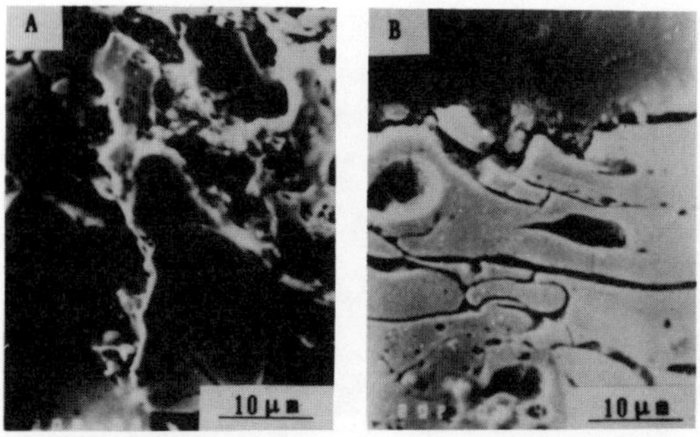

Figure 8 SEM micrographs of WC-Co coating:(A) median cracks and fractures on the worn surface and (B) cracks and fractures from cross sectional view

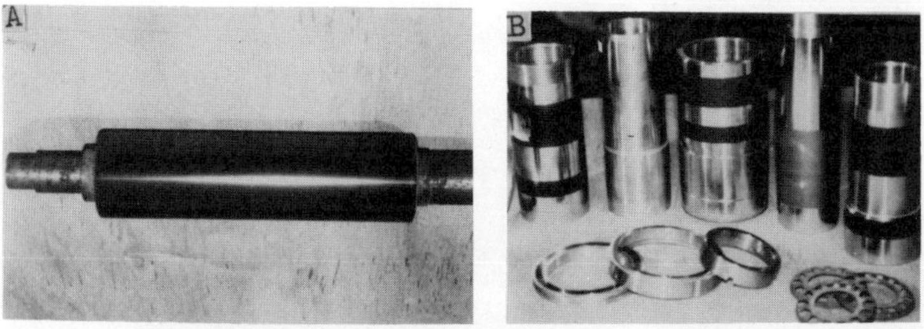

Figure 9 Some mechanical parts coated with ceramic coatings for wear resistance: (A) roller and (B) sealing ring, pump sleeves and guide rolls

3.3. Plasma Spayed Bioceramic Coatings

New materials in the field of medical implants are the result of systematic research and practical application of plasma sprayed porous coating with bio-compatibility. Oxide, bioglass and titanium coatings are being used for the surface of medical implants in which tissue growth and adherence are improved. Fig. 10 shows the SEM photo micrographs of oxide coating implanted in dogs. It has been found that the surface is surrounded with tissue of dogs and the tissue is closely attached or penetrated into the rough surface of the oxide coatings. The oxide coatings have been applied to the metallic bones and joints shown in Fig.11. Fig.12 shows the artificial thighbone coated with ZrO_2 coating. It can be observed that the tissue of the human body closely adheres to the surface of ZrO_2 coating. This might indicate that the fibers of the muscular tissue did in fact produce a mechanical interlocking effect due to the granular surface of the ZrO_2 coating [7]. More than 300 cases on patients have successfully been implanted and replaced by such kinds of bone and joint. The clinical practice for 12 years indicated that the use of new implant coated with ZrO_2 and bioglass coatings improved the therapeutical effect and

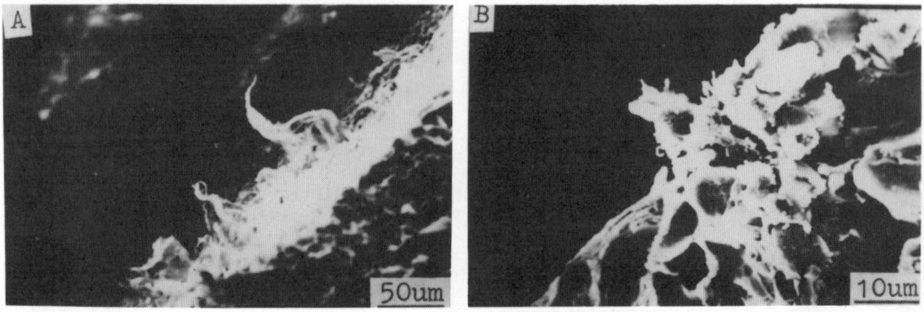

Figure 10 SEM photographs of ZrO_2 coating implanted into the leg of a dog:(A) 100 days and (B) 64 months

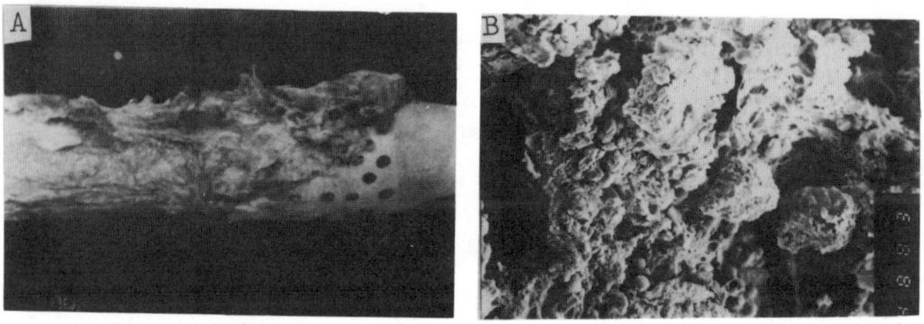

Figure 11 The thighbone coated with oxide coating after being implanted into human body: (A) thighbone (B) SEM photo

eliminated the phenomenon of sinking which always occurred in the use of metallic implants. Besides, cementless implantation was realized.

3.4. Infrared Radiation Coating

Plasma sprayed infrared radiation coatings which consist of TiO_2, ZrO_2 and Nb_2O_5 have higher emissivity at wave length ranging from 1-25 μm. The emissivity of a 75 wt%TiO_2-20wt%ZrO_2-5wt%Nb_2O_5 coating is about 0.9 as shown in Table 6. This coating has been applied onto the surface of electric heaters to improve the energy transformation efficiency [4].

Table 6. The emissivity of 75wt%TiO_2-20wt%ZrO_2-5wt% Nb_2O_5 coating

Wave band (μm)	1-25	1-14	1-8	1-4
Emissivity	0.90	0.89	0.90	0.90

***Figure 12** Various bones and joints coated with plasma sprayed ZrO_2 coating*

3.5. Diamond Film

Diamond film possesses excellent hardness, high thermal conductivity, the inertness to chemical reaction and good electronic properties and so on, which makes

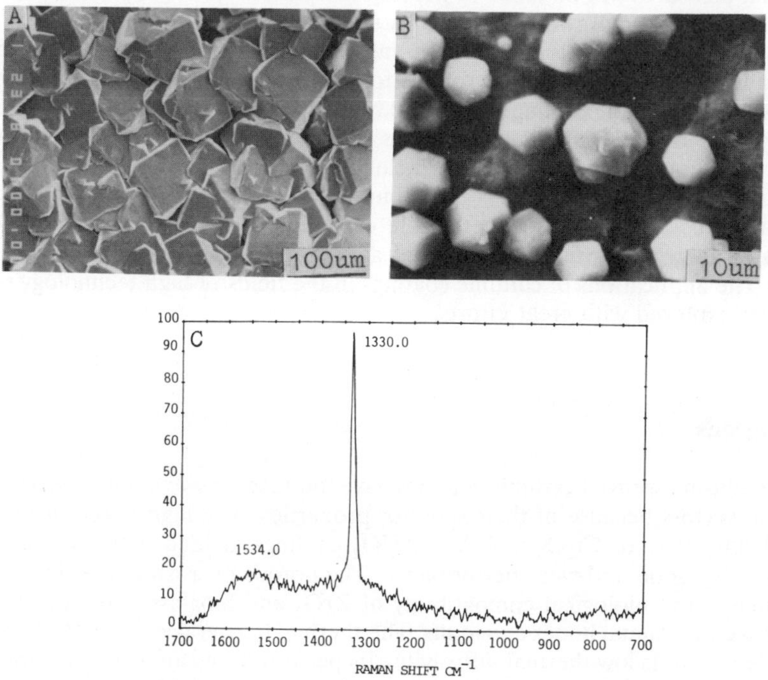

***Figure 13** SEM photo micrographs and Raman spectrum of diamond films: (A) diamond film by C_2-H_2 flame spraying; (B) diamond crystals by plasma spraying and (C) Raman spectra of a diamond film*

it a superior material for a number of applications. In the past ten years, numerous scientists have succeeded in making diamond film by low pressure deposition process[15]. More recently, thermal spraying technologies including O_2-C_2H_2 flame spraying and plasma spraying have been developed to deposit diamond films with high growth rate. Fig. 13 shows the SEM photo micrographs and Raman spectra results of the synthesized diamond films. The growth rate of the diamond film is about 60-80 μm/hr and the thickness of the diamond film is more than 0.3 mm. The diamond film produced by using a plasma spraying method is also currently being developed [16]. A report by Ohtake indicated that the growth rate of diamond film using a plasma jet can be as high as 930 μm/hr [17]. Researches indicated that the composition and flow rate of the gases, substrate material and its temperature and the pressure of deposition chamber are very important factors for the deposition of the diamond films.

4.Future Works

What measures will be taken to develop the plasma sprayed ceramic coating in the future ? Three important areas are listed as follows:

(1) New coating materials with special functions such as catalytic coating for chemical industry, new bioceramic coating for artificial bone, joint , tooth root, organ and new wear resistant coatings etc. should be explored.

(2) The mechanisms and functions of the ceramic coatings in various applications should be pursued. These include friction and wear, thermal stress and heat transfer during heating and cooling, biological compatibility, catalytic reactions on surface etc.. The effects of metastable crystalline phases, pores and defects in coatings on their properties and functions also need to be investigated.

(3) The applications of ceramic coatings in the fields of high technology should be further explored with great efforts.

5.Conclusions

The plasma sprayed ceramic coatings have found extensive applications in many industrial sectors because of their superior properties. The high hardness and high thermal diffusivity of Cr_2O_3 and Al_2O_3-TiO_2 coatings at high temperatures made them showing good anti-wear performance. The porous microstructure, inertness to environment and biological compatibility of ZrO_2 and bioglass make them more prospective than metal bones and joints. The thermal barrier coating of ZrO_2 coating has made use of its low thermal diffusivity . Its performance can be greatly improved when it combines with metal to form a graded structure. With emissivity value as high as 0.9, the infrared radiation coating consisted of ZrO_2-TiO_2-Nb_2O_5 results in large scale application in electric heaters. The plasma spraying process can also be used to deposit the diamond film. A high growth rate can be obtained. It is up to now the highest rate ever known as compared to other deposition techniques.

The lamellar structure of the ceramic coatings play an important role in wear process. The wear of the coatings happens as the result of the median crack extension.

With the improvement of plasma spraying technique, the applications of the plasma sprayed ceramic coatings will be widened, especially, in high technology fields.

5.Acknowledgements

The authors are indebted to Prof. Huang Bingtang, Senior Eng. Huiling Lin, Mr. Jianzhong Qu, Mr. Yefang Zhang and Mr. Haichao Qiao for their assistance and cooperation.

6. References

1. L. Powlowski, *Powder Metallurgy Int.* **23**(1991)357
2. C. X. Ding, B. T. Huang and H. L. Lin, in *Proc. Int Symp. on Tribochemistry*, ed. S. S. Zhang(Lanzhou Inst. Chem. Phy. 1989)308
3. Z. H. Tong, C. X. Ding and D. S. Yan, *ibid*, 299
4. C. X. Ding, N. J. Lu and H. L. Lin, *J. Inorganic Materials* **7**(1979)65 (in Chinese)
5. C. X. Ding and B.T. Huang, in *Advances in Coating Technology* Ed., E.D.Houck (ASM International,Ohio)407
6. C. X. Ding and H. C. Qiao, in *Proc. of Third Sino-Japan Symp.* on Ion Surface Optimization of Material (The Association of Ion Surface Optimization of Materials, Shanghai,1990)40
7. C. X. Ding, B.T. Huang and H. L. Lin, in *Advances in Thermal Spraying* (Pergamon Press, U.S.A.,1986)553
8. C. X. Ding, R. A. Zatorski and H. Herman, *Thin Solid Films* **118**(1984)467
9. R. W. Smith and R. Novak, *Powder Metallurgy Int.* **23**(1991)147
10. R. W. Smith and R. Novak, *ibid* **23**(1991)231
11. H. C. Qiao, J. Z. Qu, Y. F. Zhang, C. X. Ding, in *Proc. of 2nd Plasma-Technik Symp.* **Vol. 2**, Eds. S. B. Sandeir, H. Edchnauer, P. Huber and A. R. Nicoll (Hafliger Druck AG, Wettingen,1991)
12. C. X. Ding, B. T. Huang and H. L. Ling, *Thin Solid Films* **118**(1984)485
13. Z. H. Tong, C. X. Ding, D. S. Yan, *Wear* (to be published)
14. M. I. Mendelson, *Wear* **50**(1978)71
15. B. Lux and R. Haubner, *International J. of Refractory Metals & Hard Materials* **8**(1989)158
16. C. X. Ding and H.C. Qiao, *Ion Science and Technology* **191**(1991)19
17. N. Ohtake and M. Yoshikawa, *Japan Electrochem. Soc.* **137**(1990)717

Chapter 7

THE STRUCTURE OF THERMALLY SPRAYED CERAMIC COATINGS AND ITS DOMINANT EFFECT ON THE COATING PROPERTIES

Akira Ohmori and Chang - Jiu Li
Welding Research Institute, Osaka University
Osaka, Japan 567

Abstract : A thermally sprayed coating consists of flattened particles and has layer structure . An individual flattened particle is basic unit in a coating. The structure of thermally sprayed coating is a basically concerned with the structure of flattened particle including mean thickness and occurrence of microcracks in ceramic particles and bonding at the interface between flattened particles. The characterization of the structure of plasma sprayed Al_2O_3 coatings using copper electroplating technique reveals the existence of a great deal of non bonded interface area and bonding rate of 32% for the typically used coating. The pores in a thermally sprayed coating include the nonbonded interface area and vertical cracks in ceramic lamella besides commonly referred pores. The structure and property evidently illustrate that the lamellar structure of limited bonding at the interfaces between flattened particles dominates the coating properties.

1. Introduction

The properties of a thermally sprayed coating depend on its microstructure which results from certain spray process. The establishment of a rigourous relationship between properties and structure of a coating is the key point which will lead to production of a desirable coating. This essentially involves the characterization of coating structure, effect of processing parameters on the structure and correlation of property-structure relationship.

A thermally sprayed coating is deposited through the built-up of splats quenched from a series of molten droplets. An analysis of spraying process using typical spray parameters suggested that the deposition of droplets during thermal spraying is basically independent (1). Accordingly, the characterization of the structure of a coating is essentially concerned with the characterization of the structure of an individual flattened particle in the coating and the microstructure in the splat and bonding between flattened particles. the microstructure of a splat including grain size and crystal structure depends on the cooling rate after a droplet flattened. Moss (2) reported a cooling rate of 10^{7} °C from the calculation s of the heat transfer from a thermally sprayed droplet to a metal surface. This is consistent with the calculated results on cooling rate for quenched splat reported by Ruhl (3). Experimental measurements of dendrite spacing, for example, by Wilms and Herman (4) and Murakami et al (5) showed cooling rate from 10^{6} to 10^{8} °C/sec during solidification. Therefore, it is believed that the microstructure in flattened particles is preferred in meta-stable crystal structure (6) with ultra-fine grain size (7,8)? which is an intrinsic character of thermal spray process. On the other hand, the structure of a splat is related to the shape and size of the splat and simply includes the average thickness of a splat and occurrence of microcracks in the splat, in particular, in ceramic coating. Accordingly, the structure of a coating involves the structure of individual flattened particles and the bonding between flattened particles.

For engineering materials from conventional methods an ultra-fine grain size generally leads to the development of unique properties of materials (9). For mechanical properties, fine grain size corresponds to the enhancement of strength (10). However, the strength of coatings is much less than identical bulk materials (11). It is obvious that the ultra-fine grain structure contributes scarcely to the enhance the mechanical properties of coatings. On the contrary, such a fact suggests the dominant effect of structure coating including the structure of flattened particles and bonding state between flattened particles on the coating properties. This review article mainly emphasizes on the characterization of the structure of individual flattened particles and illustrates the results of some modeling recently carried out for the development of the relationship between coating structure and properties which indicate the dominant effect of limited interlamellar bonding in a coating.

2. THE STRUCTURE OF FLATTENED PARTICLES

When a droplet impacts on a substrate, the droplet generally spreads radically to form a flattened particle or splat or a flat disc. The structure of such a flattened particle including morphology and size, in particular, the mean thickness, depends basically on the state of the droplet (melting state, temperature and velocity) and the interaction with a substrate. When a sprayed particle is not completely molten, the thin partially molten outer layer spreads down to a thin layer, whereas the unmelted core of the particle may be deposited in the centre of the particles pseudodemispheric shape or may bounce off leaving a hollow center.

For a completely molten particle, it spreads out radically into a thin disc when impacting on a substrate. Practically, a flattened particle is not completely uniform in thickness as observed by Worthington (12) in the case of free falling of milk droplet on a plate and break up or break off the periphery and outer part of particles only leaves central part of flattened particle on substrate surrounded with radically stretched arms (13). A particle may also separate into several small particles (13,14). For modeling, a spray droplet impacting on a substrate is usually assumed to spread out radically and form a flat disc. The terminal thickness of flattened particle will be controlled by consumption of droplet kinetic energy against surface and viscous energies during flattening and may be affected by the solidification of droplet which may occur simultaneously with flattening.

A simple analysis of splat cooling by Jones (15) indicates that the rate of flattening of the droplet is about half the impacting velocity which is based on the numerical calculation by Harlow (16) of the flow of a water droplet impacting on a flat surface. This would be orders of magnitude larger than the solidification rate (15). The examination of flattening process of a water droplet impacting on a hot wall at velocities from 0.63 m/sec to 2.2 m/sec shown by high speed photographs by Wachters et al (17) suggests that flattening rate of a impacting droplet is little larger than half the impacting velocity and decreases with an increase in impacting velocity in the above drop velocity range. Engel (18) showed that head of water drop of 0.41 cm in size at an impacting velocity of 8.2 m/sec vanished at the end of 1.2 m/sec from high speed photographs showing flattening process. This suggests a mean flattening velocity of about 42% of the impacting velocity. In the velocity range usually observed during thermal spray process flattening rate is not available. Concerning solidification time, recently, Moreau et al (19) examine the temperature change of Nb particles of +22-45 μm in size after impacting on steel and glass substrate. The result suggested solidification time of 5 μsec and 15-20 μsec on steel and glass substrates respectively. Assuming that Nb particle has a velocity of 100 m/sec to 200 m/sec before impact and spreads to a disc of 2μm in thickness and also experiences spreading at a rate of half the impacting velocity, the flattening time for a Nb particle of mid-point particle size yields 0.3 to 0.6μsec. This analysis is consistent with that of Jones. Therefore, it is believed that flattening and solidification can be treated as two independent processes. Subsequently, the terminal thickness of flattened particle is controlled by the consumption of droplet kinetic energy against surface and viscous work. Based on the treatment of such energy originally proposed by Jones (15), following relation can be obtained:

$$d/\delta = (8\rho v d/\eta)^{0.25} \qquad (1)$$

where d = droplet diameter, v= velocity, δ= thickness of flattened particle, ρ= density of droplet, η = viscosity of liquid droplet.

Madejski treated the flattening of impacting droplet by taking the effect of solidification into consideration (20). In the case that the effect of solidification can be neglected, following relation can be obtained:

$$\frac{3\xi_m{}^2}{W_b} + \left(\frac{\xi_m}{1.2941}\right)^5 \frac{1}{R_e} = 1 \qquad (2)$$

where W (Weber number) = $\rho v^2 d/\gamma_s$

R_e(Reynolds number) = $\rho v d/\eta$

ξ_m = D/d. D = diameter of flattened particle
γ_s = surface tension of liquid

Under general thermal spray conditions, for example, in the case of alumina (21) , W_e = 1000 - 20000 and the above relation can be further simplified as follows:

$$\xi_m = 1.2941(\rho v d/\eta)^{0.2} \qquad (3)$$

By using ξ_m as a parameter, experimentally observed results showed good agreement with theoretically calculated one (31) following eqn.,(3). Practically, the occurrence of break-up at the periphery of flattened particle may make the estimation of size of flattened particle (D) difficult.

Instead, the average thickness of flattened particle would be easily available. The parameters of d/δ and D/d can be related from mass conservation as follows :

$$d/\delta = 1.5\ (D/d)^2 \qquad (4)$$

Accordingly, eqn. 3 becomes

$$d/\delta = 2.5\ (\rho v d/\eta)^{0.4} \qquad (5)$$

It can be noticed that following above treatments Jones's and Madejski's models both relate the particle flattening parameters to Reynolds number, but with different coefficients and exponents.

The effects of particle velocity on d/δ for 20 micron alumina particle at two temperatures calculated following two models (22) shown in fig. 1. the measurement of alumina lamellae in alumina coating sprayed with +10 - 44 micron powder (mean size : 20microns) yields a mean thickness of 1.75 to 2 microns. Using mean particle size yields value of 10 to 13 for d/δ . In spite of wide particle size range used in spraying, it is obvious that Madejski's model overestimates largely particle flattening. For typical molten Al droplet near melting point plasma-sprayed at a low power level with powder of -75+90 μm, an average size of 280 μm in diameter for flattened particles was observed(ref.13 and fig.2). This yields a d/δ value of 17. Table 1 shows the calculated value of d/δ for several typical particle velocities at melting point using two models. The density and viscosity are $2.39\text{x}10^3 kg/m^3$ and $4.5\text{x}10^{-3} sN/m^2$ respectively(23). Consistently, Jones's model gives reasonable value compared with observed result whereas Maddejski's model considerably overestimates the flattening of droplet. Therefore, it is reasonable that the terminal thickness of flattened particles can be estimated by Jones's model rather than by Madejski's model.

In a plasma-sprayed ceramic coating, microcracks are usually observed in individual flattened particles. The microcracks occur perpendicular to lamella in net-like form. Such vertical cracks may occur at an average interval less than 10 μm as observed for alumina coatings (24).

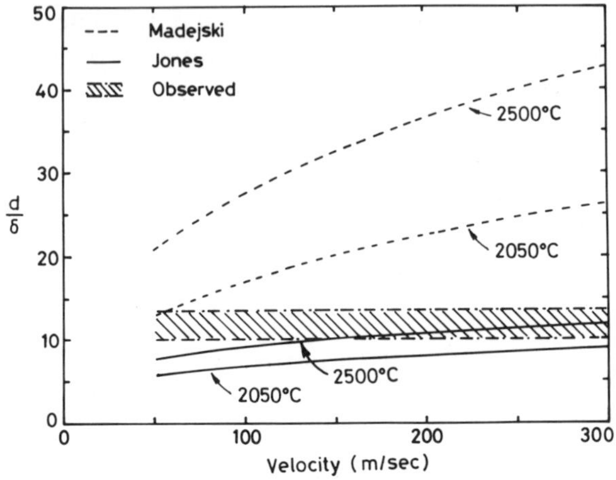

Fig.1 Comparison of the ratio of particle grain size to thickness of flattened particle estimated from Madejski and Jones's models with observed one for Al_2O_3 coating(after Ref.22)

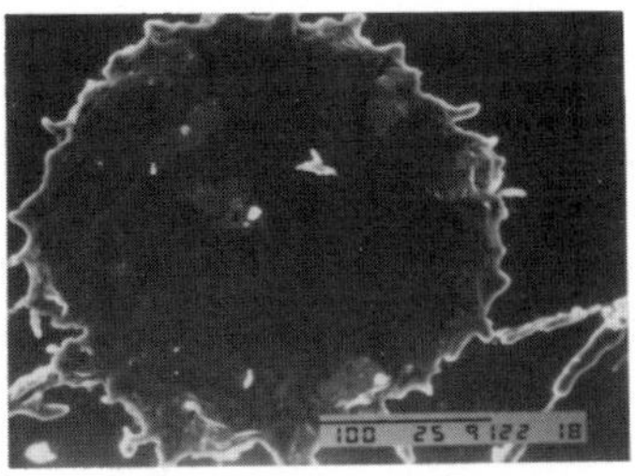

Fig.2 Plasma-sprayed typical Al splat(after Ref.13)

Table 1 Calculated d/δ using Jones and Madejski's model for several droplet velocities(after Ref.13)

d/δ \ v(m/sec)	80	100	150	200
Jones	13	13.7	15.2	16.3
Madejski	66	71.7	84.3	94.6

3. THE BONDING AT THE INTERFACE BETWEEN FLATTENED PARTICLES

The bonding at the interface between flattened particles is the most important factor in governing coating properties besides the interface bonding between the substrate and coating in substrate-coating system. the examination of fractured coating surface after fracture mechanics test (25) and shear strength test (26) suggests that cohesive fracture easily occurs and the interface area is weakest part in the coating. It is also suggested that only a limited interface area is in real contact (25). This fact was first experimentally confirmed through direct observation of plasma-sprayed alumina coating by McPhersonn and Shafer using TEM (27) and the comparison of alumina coating's Young's modulus with identical bulk material suggests a bonding rate of about one-fifth. This was further confirmed by a microstructure model for thermal conductivity of ceramic coatings proposed by McPherson (28).

Recently, electroplating copper into alumina coating revealed that metal can fill up all kinds of micro-pores including commonly referred pores , interlamellar non-bonding area, microcracks except isolated pore (29). Direct measurement of interface bonding has been achieved for alumina coatings utilizing copper plated into non-bonded interface area (24,30). A typical microstructure of copper electroplated alumina coating is shown in fig. 3 with detailed indication of structure of coating. Based on the characteristics of structure of flattened ceramic particles in the coating, such as the continuity of a cross section of a flattened particle, vertical occurrence of microcracks in individual lamella with respect to lamellar plane, the roundness of the periphery of a flattened particle, lamellar structure of a cross section of copper plated alumina coating was traced with the aid of copper plated in the coating as indicated (22) by white strings shown in fig.3. Then, cross sections of flattened particle was separated from coating cross section as schematically shown in fig. 4. Subsequent measurement shows that the interface bonding of a commonly used alumina coating sprayed under optimized conditions only reaches to 32% (30).

Figure 5 shows typical microstructure of copper plated alumina coatings sprayed at different spray distance at a plasma power of 28 kW (30). A quantitative measurement of bonding rate for above coatings shows that a rapid drop of bonding rate occurs when spray distance

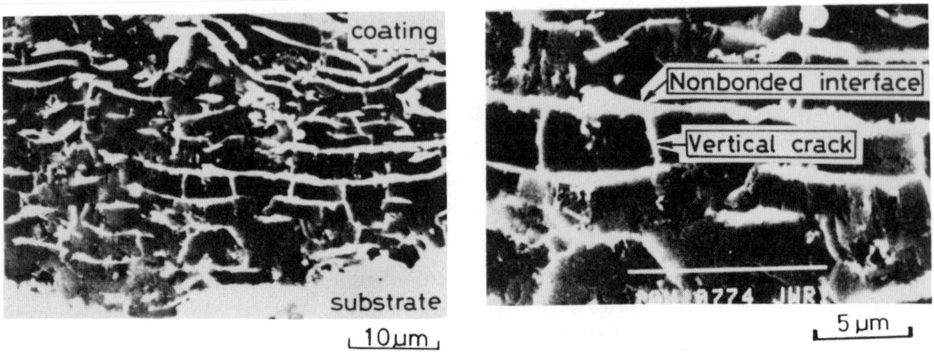

Fig. 3 SEM micrographs of copper-electroplated alumina coating. White strings are copper representing vertical cracks in individual lamella and nonbonded interfaces between flattened particles (after ref. 24).

increases from 100 mm to 150 mm as shown in fig. 6. Practically, this is a significant result.

A quantitative estimation of effect of plasma power during plasma spraying reveals that the bonding rate is rapidly saturated to 32% with an increase in plasma power (fig.7). This result implies that the increase in power of plasma spraying equipment does not contribute to an increase in interface bonding.

According to the effect of spray conditions on the particle temperature and velocity reported (31-33), it may be suggested that the bonding rate is affected by spray conditions mainly through the particle temperature rather than particle velocity (30). All those results quantitatively reveal that only a limiting bonding exists at the interfaces between flattened particles. Such results may imply the limitation of thermal spray process to produce a coating of high lamellar bonding.

The limited bonding between flattened particles was considered to likely arise from the absorbed or entrapped gases (34,35). On the contact between solid bodies, it was found that the adhesion developed between

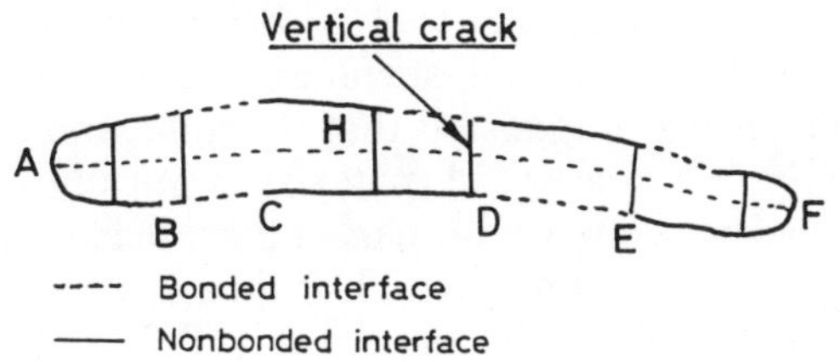

Fig.4 Schematic diagram of structure of a cross-section of a flattened particle(after Ref.24)

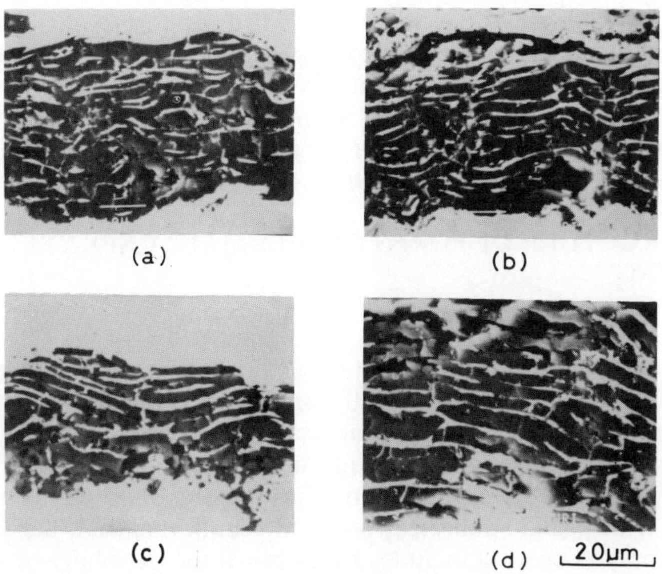

Fig.5 Typical microstructure of copper-plated Al_2O_3 coatings sprayed at different spray distance; (a) 80mm, (b) 100mm, (c) 150mm, (d) 200mm (after Ref.30)

two indium surfaces in the impact experiment (about 2 msec) is less than one-third that observed for an apparent equal area of contact, in a static test (pressed together for two minutes, ref. 36). This implies that the real contact area in the impact condition would be less than one-third of the apparent one. It was also pointed that such a difference is unlikely due to air entrapped between the colliding surfaces because experiment was carried out in vacuum. It was suggested that the adhesion depends on the contact time. Generally, when a liquid material is put onto the surface of a solid body which can be wetted with each other, it is observed that the contact angle decreases with time, which happens more fast with the increase in temperature of liquid or solid (10,36). During plasma spraying, it can be estimated that the time from impact of droplet to complete solidification is less than milliseconds based on splat cooling. The direct measurement of temperature change of spray droplet impacting on steel and glass substrate by Moreau et al (19) shows that the time from impact to complete solidification is only several or several tens of microseconds depending on the substrate. The contact time between flattened liquid and formerly formed cold coating surface is too short for liquid to wet sufficiently identical solid materials at rather low temperature. That is, dynamic process would play important role in determining interface bonding during plasma spray. Therefore, it is considered that low temperature of formerly deposited coating and short contact time is much responsible for the low bonding between flattened particles besides entrapped gases (30).

4. THE STRUCTURE OF PORES IN THERMALLY SPRAYED COATING

The porosity is usually used as a parameter of the structure of thermally sprayed coating as for other porous materials, because the porosity is usually empirically well correlated to the properties of bulk porous materials. However, the structure of pores in a thermally sprayed coating is remarkably different in morphology from those in bulk materials produced by conventional processing method as sintering. Compared with pseudo-spherical pores in a bulk material, the pores in a thermally sprayed coating includes the pores at the periphery of flattened particles which resulted from insufficient filling of subsequent droplet, nonbonded interface area and microgaps from vertical cracks in ceramic coating as shown in above section. Such kinds of pores, in particular, nonbonded interface area and microcracks, are much large in one or two directions but very small in other directions. The measurement of pore

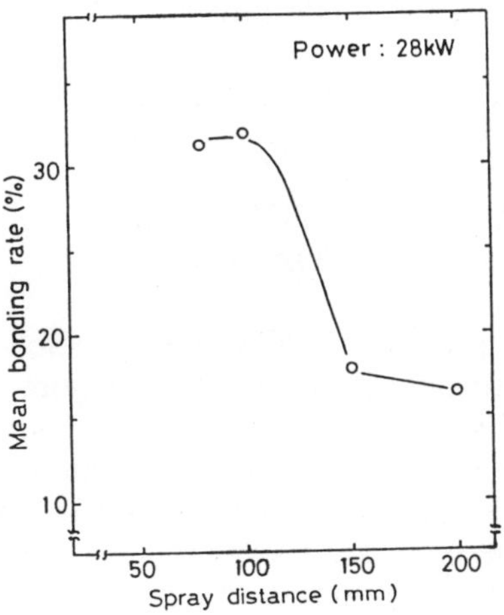

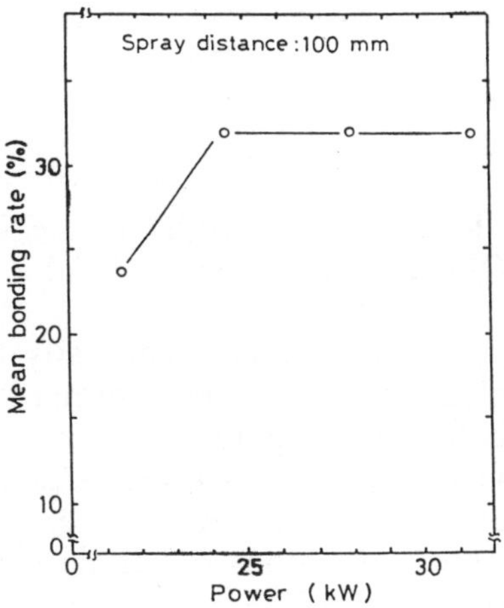

Fig.6. (left) Effect of spray distance on the bonding rate of plasma sprayed alumina coatings (after ref.30)
Fig. 7. (right) Effect of plasma power on the bonding rate of plasma sprayed alumina coatings

size in a thermally sprayed coating by mercury intrusion porosimetry (MIP) shows a bimode distribution (27,37-39). Large pores have a size from 1 to over 10 μm, but small pores in submicron size (38,39). According to the microstructure of cross section of copper plated alumina coatings, it is obvious that the small pores revealed by MIP correspond to the width of nonbonded interface area and vertical cracks, while large pores correspond to commonly referred those pores at the periphery which result from insufficient flattening of incompletely melted particles. Vertical microcracks provide many channels to interconnect nonbonded interface area, thus pores. However, small pores constituted of nonbonded interface area can not be observed from optical microstructure of a cross section (30,40). Large pores observed from a microstructure of a cross section mainly resulted from pull-out of flattened particles owing to the existence of much nonbonded interface area and microcracks (30,40,41). Therefore, much ambiguity may arise when one attempts to estimate porosity from a microstructure of a cross section of a coating. It may be more meaning ful and necessary to characterize the pore distribution rather than porosity when porosity is estimated quantitatively. Moreover, dimensions of nonbonded interface area and vertical cracks in a coating are two structural parameters, which may be estimated from pore size distribution.

5. THE RELATIONSHIP BETWEEN THE STRUCTURE AND PROPERTIES OF THERMALLY SPRAYED COATING

For porous materials, generally the properties of a coating such as mechanical properties and thermal properties are correlated to porosity empirically (42). There are many available empirical formulae which describe well property-porosity relation for materials having randomly distributed spherical pores. For example, Young's modulus and thermal conductivity of porous materials can be expressed (42) in exponential forms such as $E = E_o \exp(-bp)$. E_o is Young's modulus of dense material, p=porosity, b= constant. However, the conventional empirical relationship can not applied to thermally sprayed coating satisfactorily. In the case of Young's modulus, modulus of a thermally sprayed coating is observed to be 10 to about 30% of identical dense material (43) despite the porosity in a typical coating less than 20% from which Young's modulus value larger than 60% of identical dense material can be expected from empirical relation for bulk materials. Accordingly, in the case of sprayed porous coating,value estimated from conventional empirical relation for spherical pores is much higher than usually observed result at a similar porosity. This implies that the structure of pores in the coatings, different from those in bulk materials produced by conventional processing methods should be taken into consideration at least because the shape of pores largely affects the properties (44).

With un understanding of detailed coating structure, in particular, lamellar bonding, the relationships between properties and structural parameters evidently explain satisfactorily the dependency of properties on the structure rather than porosity.

5.1 Thermal conductivity

McPherson (28) has proposed a structural model for thermal conductivity of thermally sprayed coating as shown in fig. 8. Thermal conductivity is related to structural parameters using thermal contact resistance. For a basic cell including one bonded area in a coating of lamellar structure , total thermal resistance R_t is

$$R_t = R_c + R_b = \frac{1}{2a\lambda} + \frac{\varrho\delta}{S} = \frac{\left(1+\frac{2a\delta}{S}\right)}{2a\lambda} \tag{6}$$

where R_t=thermal contact resistivity
R_b= thermal resistance of basic cell in the direction perpendicular to lamella
R_c= thermal contact resistance (fig.9)
S = apparent area of cross section of basic cell
a = radius of contact area
λ = thermal conductivity of lamella
ρ = thermal resistivity.

Thermal resistivity of the coating is given by

$$\varrho_c = \frac{R_t S}{\delta} = \frac{\left(1+\frac{2a\delta}{S}\right)S}{2a\lambda\delta} \tag{7}$$

Noticing bonding rate between lamellae $\alpha = \pi a^2/S$, apparent thermal conductivity becomes

$$\lambda_c = \frac{1}{\varrho_c} = \frac{\frac{2}{\pi}\cdot\alpha\cdot\frac{\delta}{a}\cdot\lambda}{1+\frac{2\alpha}{\pi}\cdot\frac{\delta}{a}}$$

$$\doteq \frac{2}{\pi}\cdot\alpha\cdot\frac{\delta}{a}\cdot\lambda \tag{8}$$

$$\frac{\lambda_c}{\lambda} \doteq \frac{2}{\pi}\cdot\alpha\cdot\frac{\delta}{a} \tag{9}$$

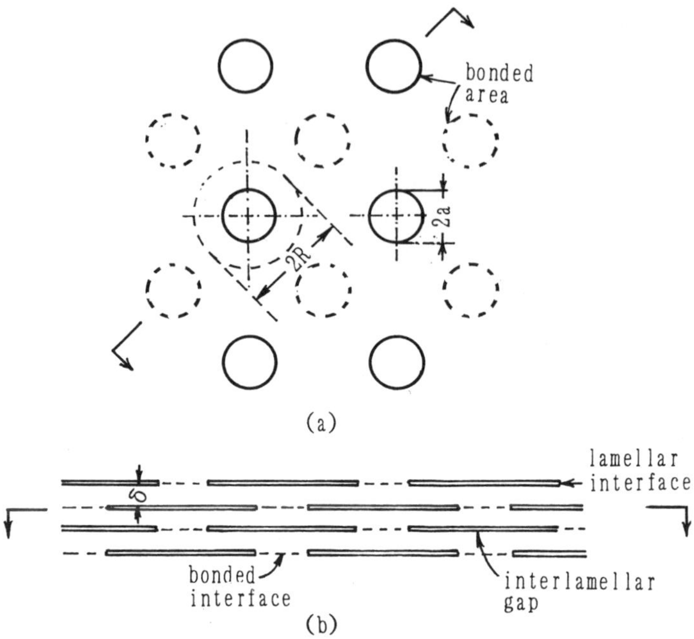

Fig.8 Schematic diagram of structural model of a thermally sprayed coating; (a) plane view of a lamella from a lamellar interface, (b) cross-section (after Ref.58)

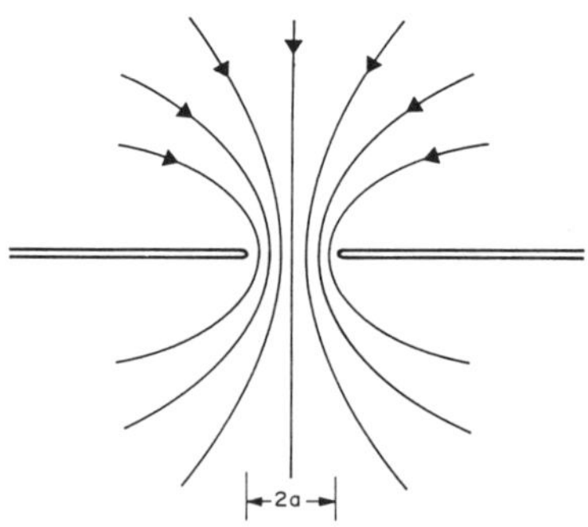

Fig.9 Heat flux through a circular true contact of radius a located within lamellar interface (after Ref.28)

It is obvious that the thermal conductivity also depends on geometrical dimension of bonding. For alumina coating, typically structural parameters are α=0.32, (δ/a) =0.5. This yields (λ_c/λ)= 0.102.

For bulk dense alumina, thermal conductivity can be found to range from 29 to 36W/mK at 300K (45-47). The thermal conductivity of alumina coating is 2.72 to 3.64W/mK (11,48). This yields a value of (λ_c/λ)= 0.076 to 0.126, which is consistent with the above estimated value although the coating consists of $\gamma-Al_2O_3$ (whose thermal conductivity is unknown) and a little higher thermal conductivity was observed by Ault for alumina coating (49).

When there are gases in the pores such as microcracks and nonbonded interface area and temperature rise up, the heat transfer from convection of gases and radiation between lamellae will increase the apparent effective thermal conductivity (50). In such a case detailed dimension of nonbonded interface area and vertical cracks will be involved (28).

It should be noticed that the above treatment on the thermal conductivity by thermal contact resistance was originally derived for electric contact resistance (51). Therefore, it is considered that above relation is also valid for electric conductivity.

5.2 Fracture Toughness

For bulk materials, critical kinetic strain energy release rate G_{1c} is generally expressed as

$$G_{1c} = 2\gamma_e \tag{10}$$

where γ_e is effective surface energy.
For cohesive fracture of coating, G_{1c} can be simply expressed by the following equation (22)

$$G_{1c}=2C_p\gamma_e\alpha \tag{11}$$

where C is a fracture path related constant, which depends on flatness of flattened particle in thermally sprayed coating. For a coating consisting of well flattened particles, C_p =1. A plasma sprayed coating can be considered to belong to such a case.

In the case of bulk alumina, γ_e varies from 25J/m^2 to 45 J/m^2 as grain size decreases from about 20 μm to less than 2 μm (53). This yields a G_{1c} of 90J/m^2 for bulk alumina with fine grain structure. Regarding the fine structure observed for alumina coatings and typical lamellar bonding rate, a G_{1c} value of 29J/m^2 can be expected for α-Al_2O_3 coating from eqn.11.

A G_{1c} of 12-26J/m^2 was observed for Al_2O_3 coating measured by DCB method and G_{1c} also depends on powder particle size distribution (54). However, Al_2O_3 coating consists of mainly of γ-Al_2O_3 . Although the data of effective surface energy of γ-Al_2O_3 is not available, it may be expected to be lower than that of γ-Al_2O_3 . Accordingly, reported G_{1c} is reasonably consistent with the expected one. This fact evidently suggests that the fracture toughness of a coating is mainly governed by interface bonding between flattened particles.

5.3 Particle impact erosion

For a ceramic coating of lamellar structure with limited bonding at the interface between flattened particles, it was found that the erosion of coating occurs through separation of flattened particles from interface after impact by erosive particle, in particular, at 90° impact (55,56). If one assumes that partially absorbed kinetic energy of impacting particle by coating contributes the debonding of bonded interface area of surface lamella exposed directly to impact, which leads to the separation of surface lamella from coating, the erosion rate can be related to the structural parameters by following equation

$$\text{ACT-JP value} = 2\,\gamma\alpha / KE_m\delta \qquad (12)$$

where ACT-JP value is reciprocal of erosion rate, E_m is kinetic energy of impacting particle, K is effective energy absorbing coefficient.

Figure 10 shows observed relation between ACT-JP value and interface bonding rate between flattened particles for alumina coatings.

Erosion test was carried by ACT-JP (Arata Coating test with jet particles) erosive tester at impacting degree of 90 with fused alumina grits of 240-420 μm at a calculated particle velocity of 63m/sec for a mean particle size of 330 μm. Experimentally, it is confirmed that particle erosion of a ceramic coating will be governed by interface bonding between flattened particles.

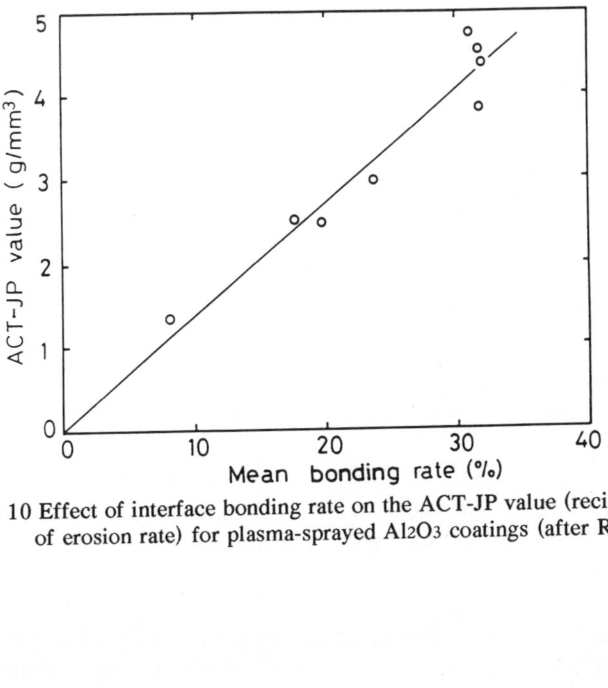

Fig. 10 Effect of interface bonding rate on the ACT-JP value (reciprocal of erosion rate) for plasma-sprayed Al_2O_3 coatings (after Ref.57)

5.4 Young's modulus

When a load is applied to a coating perpendicular to the plane of coating, the stress will be transferred from one lamella to the other through bonded interface area. Figure 11 shows schematically the loading at a cross-section of one lamella (58). Under such stressed condition, the lamella will experience tensilon at bonded interface area and bending at non bonded area as well. By applying plate theory to the coating, apparent Young's modulus of thermally sprayed coating (E_c) can be expressed as

$$\frac{E_c}{E} = \frac{\alpha}{1 + 2\pi \left(\frac{a}{\delta}\right)^4 \beta^2 \left[f_1(\beta) - f_2(\beta)\right]} \qquad (\alpha < 40\%) \tag{13}$$

$$\frac{E_c}{E} = \alpha \qquad (\alpha \geq 40\%) \tag{14}$$

$$f_1(\beta) = \frac{3(1 - \nu^2)}{2\pi}\left[\left(1 + \frac{1}{2}\frac{1-\nu}{1+\nu}\right)\left(\frac{\beta^2 - 1}{\beta^2}\right) + \frac{1+\nu}{1-\nu}\frac{2}{\beta^2 - 1}\ln^2 \beta\right]$$

$$f_2(\beta) = \frac{3}{2\pi}\frac{\left[(1-\nu)(\beta^2 - 1) + 2(1+\nu)\beta^2 \ln \beta\right]^2}{\beta^2(\beta^2 - 1)\left[(1+\nu)\beta^2 + (1-\nu)\right]}$$

$$\beta = \frac{R}{a}$$

E = Young's modulus of lamella material
ν = Poisson's ratio

Figure 12 shows the relative Young's modulus of coatings against bonding rate at different a/δ values. Regarding the limited interface bonding , above model quantitatively explains usually observed Young's modulus of thermally sprayed coating.

The measurement also shows anisotropy in Young's modulus. Young's modulus in direction parallel to lamellar plane is larger than that in the perpendicular direction.

All above results evidently show lamellar structure of limited bonding at lamellar interfaces controls the properties. The properties of a coating are also affected by geometrical dimension of bonded area besides total bonding rate, in particular, ratio of lamellar thickness to size of bonded area. the understanding of coating structure and structure-property relations will greatly contribute to the understanding of formation

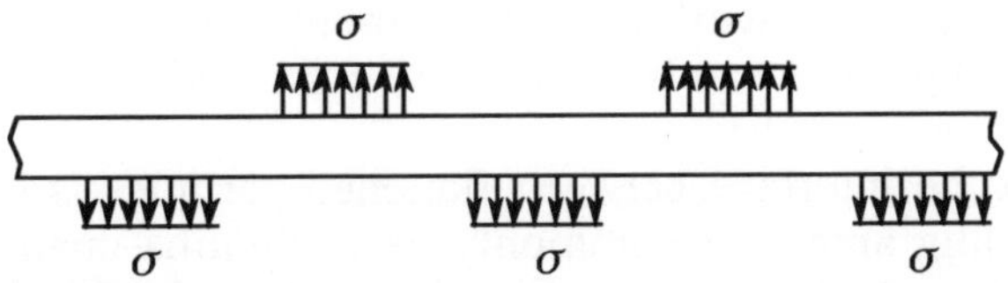

Fig.11 Schematic diagram of loading at a cross-section of one lamella(after Ref.58)

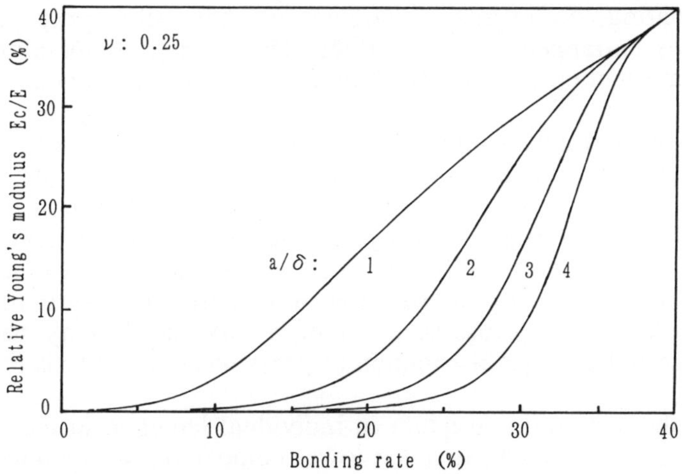

Fig.12 Effect of structural parameters on the relative Young's modulus of a thermally sprayed coating (after Ref.58)

mechanism of thermally sprayed coatings and to the further development of thermal spray technology.

6. SUMMARY

A thermally sprayed coating is characterized by layer structure. A flattened particle is a basic unit in a coating. The structure of a coating is basically concerned with the structure of individual flattened particles in the coating and the bonding between those particles. The thickness of flattened particles and interface bonding are two important structural parameters. It is illustrated that Jones's treatment on flattening of a molten droplet correlates well the experimental results. Terminal thickness of flattened particle can be reasonably estimated.

The bonding at the interface between flattened particles is much less than apparent bonding area. For commonly used alumina coating, using copper electroplating technique, the bonding rate of 32% has been estimated. The spray parameter, in particular, spray distance will largely affect the interface bonding. Such a low interface bonding rate is believed to result from limited available time for liquid droplet to wet deposited identical coating, an intrinsic character of thermally spray process, besides gases entrapped at the interface. The particle temperature may have a larger effect on the interface bonding than particle velocity.

Pores in a thermally sprayed coating consist of commonly referred large pores of 1μm to over 10 μm, nonbonded interface area and vertical cracks in individual ceramic lamella which are in submicron size range. The limited poor interface bonding and occurrence of vertical cracks make pull-out easily occur during preparation of sample for examination of microstructure. This causes the ambiguity of evaluation o\ microstructure from a cross-seection of a coating. It may be more meaningful to indicate the distribution of pores when porosity is referred.

Despite the fine grain structure in individual lamellae in the coating the properties of a ccoating is much lower compared with identical bulk material of same porosity level. Although available data are much limited and difficulty arises using alumina coating, which is different in crystal structure from bulk one, attempts on the modeling of arelationship between properties and structural parameters evidently illustrate that the lamellar structure and the limited interface bonding are much responsible for low coating properties. The understanding of detailed coating structure and structure-property relations will greatly contribute to the understanding of formation mechanism of a thermally sprayed coatings and to the further development of thermally spray technology.

The limited interface bonding between flattened particles would restrict the development of superior properties from ultra - fine grain structure in individual lamellae. It may be expected that the properties which are not affected by interface bonding can be exploited to its maximum from view point of ultra - fine grain structure.

References

1. J. M. Houben and G. C. Liempd, Proc. 10th Int. Thermal Spraying Conf., Essen, May 1983, German Welding Society, 1983, p46.
2. M. Moss, Acta Metall. 16 (1968) 321
3. R. C. Ruhl, Mater. Sci. Eng., 1 (1967) 313
4. V. Wilms and H. Hermann, Proc. 8th Int. Thermal Spraying Conf., Florida, American Welding Society, 1977, p236
5. K. Murakami et al. Proc. 3rd Int. Thermal Spraying Conf., California, ASM Intl. 1991, p 351
6. R. McPherson, J. Mater. Sci., 8 (1973) 851
7. S. Safai and H. Herman in Treatise on Materials Sciene and Technology, vol. 20 , ed. H. Herman (Academic Press) 1981, p183
8. R. McPherson, Proc. Int. Symp. Ad. Th. Spray. Tech. and Allied Coatings, Osaka, Japan High Temp. Soc., 1988, p24
9. H. Jones J. Mater. Sci., 19 (1984) 1043
10. E. D. Kingery et al. Introduction to Ceramics (Wiley Interscience, 1977)
11. Japan Thermal Spraying Society, thermal spray handbook, 1984(in Japanese)
12. A. W. Worthington, Proc. Roy. Soc. 25 (1877) 261; 25 (1877) 498
13. C.J.Li and A. Ohmori, unpublished
14. J. Zaat, Ann. Rev. Mater. Sci., 13 (1983) 9
15. H. Jones, J. Phys. D. Appl. Phys. 4 (1971) 1657
16. F. H. Harlow and J. P. Shannon, J. Appl. Phsy. 38 (1967) 3855
17. L. H. J. Wachters and N.A. J. Westering, Chem. Eng. Sci., 21 (1966) 1047
18. O. G. Engel, J. Res. Nat. Bureau of Standards, 54 (1955) 281
19. C. Moreau et al. Surface Coating Tech., 46 (1991) 173
20. Madejski, Int. J. Heat and Mass Transfer, 19 (1976) 1009
21. R. McPherson ,J. Mater. Sci. 15 (1980) 3141
22. C.-J. Li Doctoral thesis, Osaka University (1989)
23. Metal Handbook: ed. H. Boyer and T. L. Gall, ASM, 1984
24. A. Ohmori and C.-J. Li , Thin Solid Films, 201 (1991) 241
25. C. C. Berndt and R. McPherson, Proc. 9th Int. Thermal Spraying Conf., Hague, 1980, Nederlands Inst. voor Lastetechnik, (1980) 310
26. L. W. Crane et al. in ref. 1 p 46
27. R. McPherson and B. W. Shafer, Thin Solid Films, 97 (1982) 201
28. R. McPherson, Thin Solid Films, 112 (1984) 89
29. Y. Arata et al. Trans. Jpn. Weld. Res. Inst. 16 (1987) 259
30. A. Ohmori et al. Trans. Jpn. Weld. Res. Inst. 19 (1990)259

31. A. Vardelle et al. ref 25, p155
32. M. Vardelle et al. ref. 1, p88
33. P. Fauchais et al. Met. Trans. 20 B (1989) 264
34. T. Gillespie and S. E. Rideal, J. Colloid. Sci. 10 (1955) 281
35. R. McPherson, Thin Solid Films, 83 (1981) 297
36. A. W. Crook and W. Hirst, Research, 3 (1950) 432; T. Yamaguchi et al. in Surface and Interfaces in Ceramic and Ceramic-Metal System, ed. J. Pask and A. Evans, 1981,p503
37. R. McPherson and P. Cheang, Proc. 12th Int. Thermal Spraying Conf., London, june 1989, p17-1
38. R. McPherson and P. Cheang, in High Performance Ceramic Films and Coatings, ed. P. Vincensini (Elsevier, Netherlands, 1991) p 277
39. S. Uematsu et al. ref. 8. p379
40. V. S. Thompson and O. J. Whittemore, Ceramic Bull., 47 (1986) 637
41. M. F. Smith et al. Proc. 4th National Spraying Conf., Pittsburgh, 1991, ASM inst., 1992, p 97
42. R. W. Rice, in Treatise on Materials Science and Technology, Vol. 11, ed. R. K. Maccrone (Academic Press, 1977) p 199
43. S. Kuroda and T. W. Clyne, Thin Solid Films, 200 (1991) 49
44. J. D. Eshelby, Proc. Roy. Soc. A 252 (1959) 561
45. J. Francel and W. D. Kingrey, J. Am. Ceram. Soc., 37 (1954) 80
46. Thermophysical Properties Research Center, Data Book, 3 (1966)
47. K. Tani et al. Proc. 13th Int. Thermal spraying Conf., Florida, 1992
48. Data book of Japanesse Socieety of Mechanical Eng., Heat Transfer, 1986
49. N. N. Ault, J. Am. Ceram. Soc., 40 (1957) 69
50. D. Matejka and B. Benko, Plasma Spraying of Metallic and Ceramic Materials (John Wiley, 1989) p 51
51. F. P. Bowden and D Tabor, Friction and Lubrication of Solids (Oxford Press, 1964) p 25
52. R. B. Lawn and T. W. Wilshaw, Fracture of Brittle solids(Cambridge University Press) p 65
53. L. A. Simpson , in Fracture Mechanics, ed. R. C. Bradt et al (Plenum, NY) p 567
54. P. Ostonjc and R. McPherson, ref. 37, p 9
55. Y. Arata et al., Trans. Jpn. Weld. Soc., 15 (1986) 339
56. R. Kingswell eet al., ref. 5, p 179
57. Y. Arata et al., J. Jpn. High Temp. Soc., 14 (1988) 220 (in japanese)
58. C.-J. Li et al., Proc. Austceram'92, Melbourne, Aug 1992

Chapter 8

HIGH VELOCITY OXY-FUEL SPRAYING - A REPLACEMENT FOR PLASMA SPRAYING IN SOME AREAS?

A.R.NICOLL, A. BACHMANN, G. BARBEZAT, J.R. MOENS
PLASMA - TECHNIK AG, WOHLEN, SWITZERLAND

Abstract

Plasma spraying as an industrial process producing high quality coatings is used in many applications. However, there are some areas where high velocity oxy-fuel processes are proving to be very successfull. This paper considers the use of high velocity oxy-fuel processes and discusses the coatings available for applications based on deposition using the CDS process. Coating development directions are reviewed briefly.

Introduction

The last few years have seen a tremendous increase in the introduction of high velocity combustion or oxy-fuel spraying. Explained very briefly, the reason for the success is firstly, the fact that the process is based on high velocity and not on high temperature, thus making it very suitable for carbide containing coatings and secondly that thick (>1mm) adherent coatings can be produced thus opening up the area of repair.

Originally, this market was the single domain of coatings produced by the Union Carbide D-gun or detonation process, a proprietary method of depositing coatings. This changed with the introduction of the original JET-KOTE process by Jim Browning. Unfortunately, it took many years to bring a uniform system to the market place and subsequently, many different Jet-Kote modified machines are in use today.

Today, alternatives to Jet-Kote have reached the market place e.g. CDS / 1-4/, Diamond Jet, Top Gun etc. and an alternative to the D-gun can also be obtained.

As mentioned above,in the past the units presented to the market represented stand-alone devices which did not take into account the absolute necessity of integrating the components of the system or of applying them to an on-line production scheme using robotics and manipulators. This is discussed below in the section Man-Machine-Interface.

Typically in thermal spraying, a powder is heated and accelerated towards a cleaned and grit-blasted substrate where it impacts and interlocks with the substrate. The relative movement between the spray gun and the component results in the deposition of the coating. Unfortunately, the deposited coating can consist of different components such as oxides, porosity and unmelted particles. In the CDS gun, a gas mixture is formed and combusted, this being oxygen and a fuel gas, which is propane, propylene or hydrogen depending on the availablity of gas in the world and the type of coating required. A schematic is shown in Fig. 1.

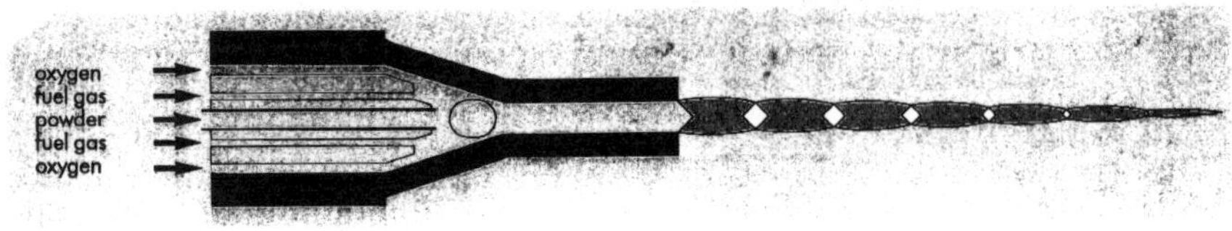

Fig. 1 Cross-section of a CDS spray gun.

The advantage of the high velocity flame spray process is that it is characterised by high particle velocities and low thermal energy levels thus making it particularly advantageous for the deposition of carbide containing coatings such as tungsten carbide cobalt or chromium carbide nickel chromium. Indeed, particle velocities of 400m/s have been measured for tungsten carbide cobalt coatings deposition conditions /5/ as compared to 100-150m/s for plasma sprayed coatings.

The high quality of the carbide coatings in use today depends also on the extent of the carbide transformations that occur which in turn can be influenced by the spray process paramenters, i.e. choice of fuel gas, flow rates, powder selection /6/, length of the spray gun barrel and the use of additive gas to cool the gas stream where required. Table 1 shows the range of gas flows being used.

Oxygen l/min	420	210	420
Propane l/min	55	-	-
Propylene l/min	-	-	55
Hydrogen l/min	-	420	-
Carrier Gas l/min		10 - 30 Nitrogen	
Spray Distance mm		200 - 300	
Typical Feed Rate g/min		40 - 60	

Table 1 Typical HVOF spray conditions for oxygen with propane, propylene or hydrogen respectively.

Markets and material developments

Tungsten carbide coatings are particularly suited for protection against abrasion and erosion induced wear, as well as fretting. They are used in a temperature range up to a maximum of 500°C and chemically the carbide particles can be bonded with cobalt, nickel or cobalt-chromium, the latter combination improving the corrosion resistance of the resulting matrix. Thus, applications for carbide containing coatings are found in areas requiring wear protection such as the iron and steel industry where large rolls are sprayed for protection (Fig. 2) or, for example, the aircraft industry where compressor and fan blades of aluminium or titanium are protected against erosion or hammer wear respectively.

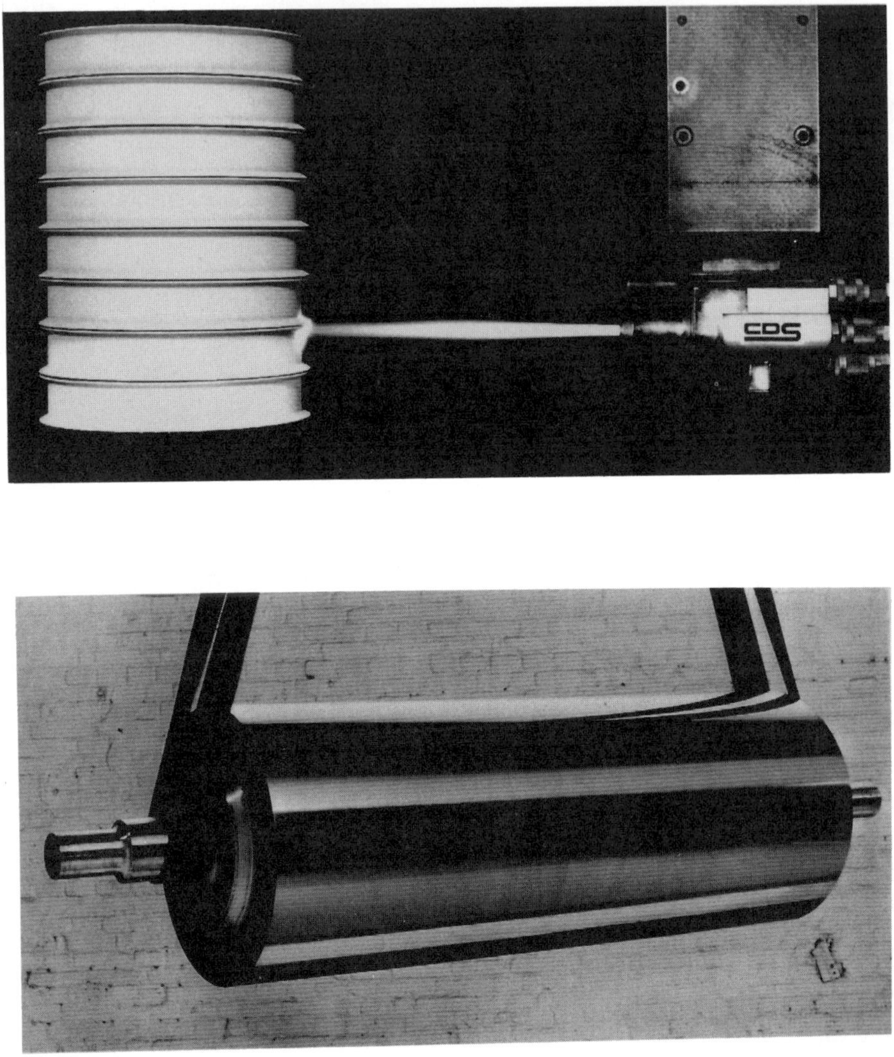

Fig. 2 Examples of CDS sprayed tungsten carbide onto components for the steel industry. (Courtesy of Advanced Coating, Belgium)

Indeed, on soft substrates such as titanium, aluminium or copper, the velocity of the tungsten carbide-cobalt particles is so high that grit-blasting of the substrate prior to coating deposition is no longer necessary. The first particles that impact on the substrate actually grit blast the surface and as the roughness increases the particles start to adhere and build-up the coating.

Demands are growing in various industries for the deposition of materials for abradable, repair and anti-fretting applications. In many cases, the driving force behind this requirement is the fact that specific components have to sprayed in different areas with variety of coatings respectively. For example, large fan blades in the aircraft industry are coated in one area with tungsten carbide cobalt and in another area with an anti-fretting coating such as copper-nickel-indium. Thus, the introduction of the process depends on the ability to spray both coatings. An example of copper-nickel-indium (Amdry CDS 4302) is shown in Fig.3 sprayed with an oxygen-propane gas mixture.

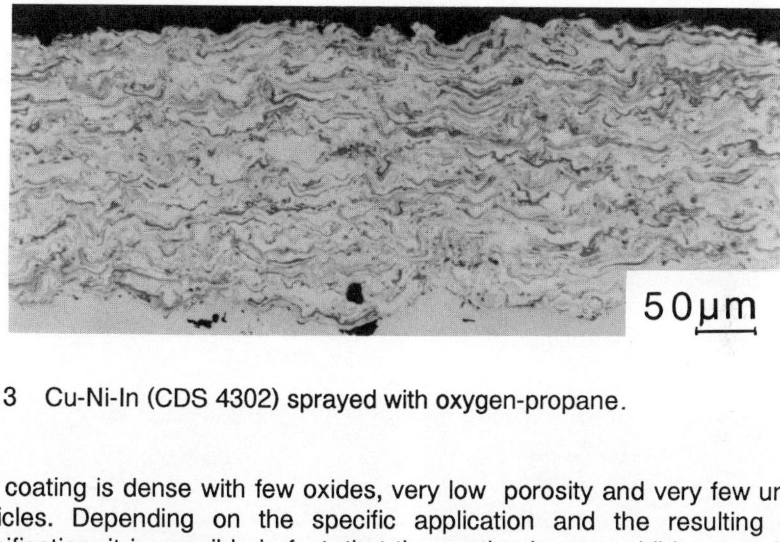

Fig. 3 Cu-Ni-In (CDS 4302) sprayed with oxygen-propane.

The coating is dense with few oxides, very low porosity and very few unmelted particles. Depending on the specific application and the resulting coating specification, it is possible in fact, that the coating has to exhibit a certain oxide content or porosity in order to maintain the correct properties e.g. compressive stresses. Even with a large number of unmelted particles it is possible to have dense coatings as shown in Fig. 4 for steel powder sprayed using propane as the fuel gas.

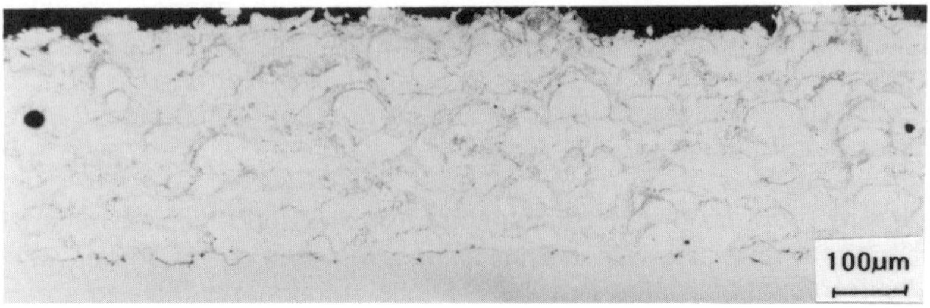

Fig. 4 HVOF sprayed steel coating.

The amount of oxide that appears in a coating can be reduced by the choice of the gas flow rates and the gases selected. For example, the material Inconel 718 (CDS 1718) is used to repair aircraft components with layer thicknesses being sprayed up to 2mm.

Important for this coating is the use of hydrogen as fuel gas to provide a reducing effect i.e. lower the oxide content during spraying. A comparison of the use of propane and hydrogen as fuel gases is shown in Fig. 5.

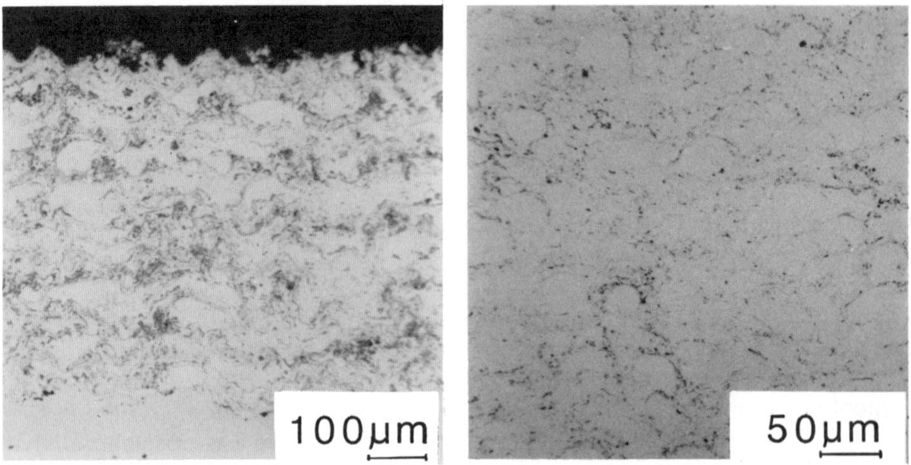

Oxygen-Propane Oxygen-Hydrogen

Fig. 5 Inconel 718 (CDS 1718) sprayed with different fuel gases.

Similarly, components for industrial gas turbines can be repaired using superalloy material such as Inconel 738 or 625 or surfaces exposed to high temperature oxidation can be sprayed with NiCrAlY materials using hydrogen fuel gas which results in very low oxide contents in the coating.

An important area for the future is gas path seals based on the use of aluminium-silicon mixed with polyester or high temperature polyimide (Fig.6). These materials are applied using the plasma spray process with either argon-hydrogen or argon-helium gas mixtures and up to approximately 70% spray deposition efficiency can be obtained. The coatings are used up to temperatures of 300 or 350°C respectively depending on the choice of polyester or polyimide.

Abradability tests have indicated that coatings applied using the CDS process have superior wear characteristics compared to plasma sprayed coatings /7/.

Applications for these coatings will be in aircraft engines for compressor seals, turbocompressors, exhaust gas turbochargers for automotive applications, industrial gas turbines and possibly steam turbines.

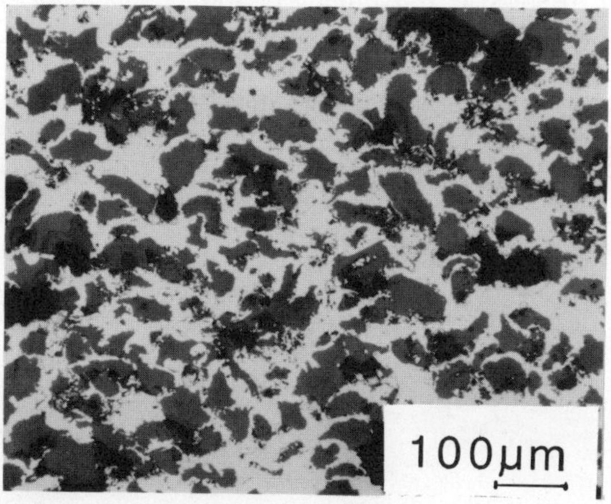

Fig. 6 CDS sprayed Aluminium-Silicon-Polyimide.

In the manufacture of glass items, moulds are typically lined with a NiCrSiB alloy to resist wear. The coating is applied using flame spraying and then melted into the substrate, this operation usually being carried out manually. The need for this is to provide bond strength and generate a coating structure with sufficient hardness and integrity to resist wear.

The same coating applied using the CDS process has sufficient bond strength and hardness that a manual heat treatment is no longer necessary thus leading to savings in the manufacture and repair of moulds. Great potential appears to exist in this area based on the ability to modify this family of alloys with additions that can raise the wear resistance considerably.

In the field of automotive applications, many components are coated with molybdenum wire using flame spraying. The resulting coating has a high hardness due to the oxidation of the molybdenum during spraying but a low interparticle coating cohesivity. Hardness in HVOF coatings as well as in plasma coatings can be generated using spray powder doped with oxygen in the form of oxide. Components requiring layers on an internal surface can only be coated individually using flame spraying and not to their final dimension. High velocity flame spraying can provide coatings of high hardness (800-900 HV0.1) and spray several components at one time at an angle due to the high velocity available (Fig.7).

The high velocity of the CDS process in spraying materials (immiscible compounds) which can easily decompose e.g. bearing materials containing copper and lead, results in retainment of the constituents and thus a new interesting application for the future.

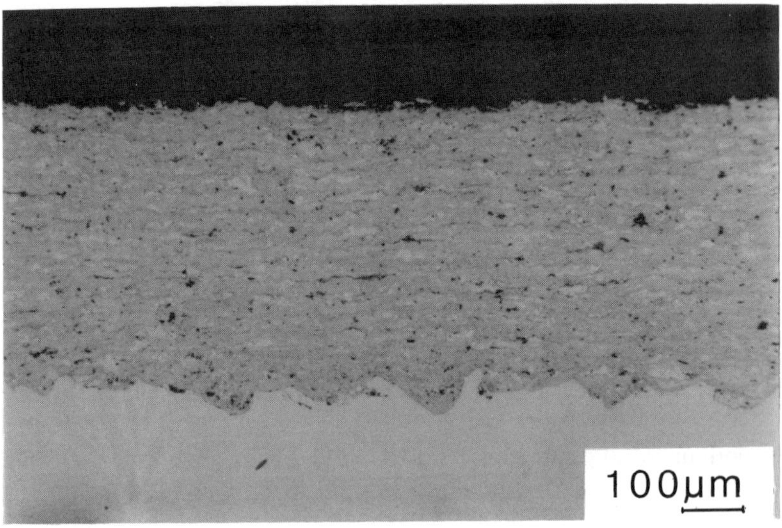

Fig. 7 CDS sprayed molybdenum doped with oxygen.

In all these industrial examples, production demands strict levels of operating system repeatibility and coating quality control in order to ensure overall coating solution reproducibility on a production basis.

Man-Machine Interface

Integration of the deposition process into a production facility requires that various demands are met, the most important of which concerns the interface between the operator and the coating machine. Certain measures have to be taken to ensure that the operator can work satisfactorily over long periods of time whilst maintaining the quality of the coating.

The process takes place inside a sound-proof booth which protects the operator from the noise and overspray, the latter requiring the use of an efficient exhaust system meeting national guidelines for emissions together with a fresh-air supply. However, placing the spray system inside a sound proof cabin requires that the system can be started satisfactorily from outside of the cabin. This requires features such as automatic gun ignition, preprogrammed start-up and stop cycles including ramping of the gases to full flow conditions. The operator has to ensure that the powder has been placed in the powder feeder located outside of the spray cabin and set the gas flow conditions on the control console in litres/minute. Together with operating system reliability, this means that attention is paid to all parts of the system such as gas supply, choice of fuel gas, powder size distribution, parameter control over long time periods and coating properties. For example, in the case of gas flow control very precise and reproducible gas flow is required that can only be guaranteed using mass flow control with closed loop regulation. Gas controls of this type can be calibrated prior to service and checked or recalibrated as required. Mass flow control makes parameter transfer possible from spray system to system and from location to location. It has to be noted that rotameter systems are subject to wear and cannot be recalibrated.

The coating machine is typically a spray gun mounted on a movement device such as a robot which moves the spray gun relative to the component to be sprayed which is mounted on a manipulator system.

Processing of a part requires mounting the component on the manipulator, closing the spray cabin doors and igniting the spray gun from outside the spray cabin.

It is clear from some of the examples above that different industries have different shapes, sizes and numbers of components that have to be sprayed meaning that the optimised and most efficient manipulation system requires careful analysis and selection and that a universal solution is generally not possible.

Fig. 8 CDS production spray facility.

In addition to stand-alone systems, the introduction of CDS has been combined with the advantages of a fully developed and equipped plasma system and in this way it is possible to combine the advantages of plasma spraying with those of HVOF in the form of CDS.

Conclusions

High Velocity oxy-fuel spraying has attained commercial acceptance with regard to coating quality. The step now being taken concerns the introduction of complete spray devices adapted to the high level production schemes previously seen only in connection with plasma spraying. In this respect, greater attention is being shown to the man-machine interface and to the introduction of new coating concepts and applications that have to be adapted as quickly as possible to high volume production.

References

/1/ G.Barbezat, High Tech. Materials and Finishing (1989),Berlin, 12-14 March 1989.

/2/ G.Barbezat, J.R.Moens, Sulzer Technical Review 1/1990

/3/ J.R.Moens, A.R.Nicoll, CDS, Kolloquim Hochgeschwindigkeits-Flammspritzen, Munich,16.11.1989.

/4/ G.Barbezat, J.R.Moens, S.Siegmann, Technischen Rundschau Sulzer 4/1990, Hochtemperatur-Supraleiterschichten : Erste technische Anwendung.

/5/ M.F.Smith, 1st. Plasma Technik Symposium Lucerne, 18-20,May 1988,Vol.1,pp77-85

/6/ G.Barbezat, J.R.Moens, A.R.Nicoll, Sulzer Technical Review 2/1991, Properties and Applications of CDS Coatings.

/7/ B.Miller, R.Schmid, A.R.Nicoll,Conf. Proc. NTSC Long Beach/USA,1990, Thermal Spray Research and Applications, ASM Int., 1991, p121.

Chapter 9

PLASMA SPRAY APPLICATIONS FOR ELECTRONIC MATERIALS

R. Suryanarayanan
Laboratoire de Physique des Solides de Bellevue
CNRS, 92195 Meudon, France

Plasma spray applications have been mostly directed to disciplines such as chemical, mechanical and metallurgical engineering. However, quite recently the plasma spray technique seems to capture the attention of another branch, namely electronic materials. This paper reviews the utilisation of plasma spraying in the preparation of several electronic materials such as semiconductor Si, Sm-Co permanent magnets, the newly discovered superconductors and amorphous metals. Methods of improving the electronic properties and some applications are pointed out.

1. Introduction

Plasma spraying is a well established technique in vogue for more than half a century for preparing a wide variety of metals, intermetallics, oxides and cermets. The applications have been mostly directed to disciplines such as chemical, mechanical and metallurgical engineering. However, quite recently the plasma spray technique seems to capture the attention of another branch, namely electronic materials. The aim of the present paper is to bring forth this new area of application. We have chosen three materials differing widely in electronic properties to illustrate this view point. These are semiconducting silicon, metallic Sm-Co magnets, magnetic oxides commonly known as ferrrites and the recently discovered oxide superconductor $YBa_2Cu_3O_{6+z}$ usually denoted as YBaCuO. The possibility of preparing amorphous magnetic metals will also be indicated. Recent results on recrystallization techniques are also described. One of the attractive advantages offered by this technique is that in a single operation one can melt the material, quench it and obtain it in a short time in a desired form having large dimensions. The cost effectiveness resulting from the automation of the technique is also not to be neglected.

2. EXPERIMENTAL TECHNIQUES

2.1 Plasma spray set-up

We used a commercially available plasma torch (METCO 7 MB, U. S. A.) with a copper nozzle of type GH (diameter 7.5mm). The powder to be sprayed was injected into the plasma flame by a neutral gas such as argon using the powder distributor (METCO 3 MP). A substrate handling system suitable for laboratory use was designed in our laboratory which is described briefly here (1). Six substrates at a time can be held along the periphery of a stainless steel cylinder (diameter = 20 cm). at equidistant. This cylinder can be rotated manually. The substrate to be coated (2.5x2.5x1 cm) is fixed to a thermocouple held firmly by spring contacts. In a batch process, the temperature of the substrate can be monitored before, during and after deposition. Accurate spraying time is achieved by the use of a hollow water cooled stainless steel shutter placed in between the torch and the substrate holder. The plasma torch mounted on a carriage can be moved to and fro from the substrate and the torch-substrate distance which is an important process parameter, can be measured and eventually kept constant during a batch process. We have used different substrates such as stainless steel, copper and alumina. When we want to obtain a deposit of a new material, the process parameters such as torch-substrate distance, duration of pre and post heating, powder feed rate, plasma arc intensity and gas flow rates (argon, hydrogen) are optimized by repeating the experiment and changing the parameters until a deposit with minimum defects are obtained for which the present set-up was found to be quite useful.

2.2 Powder

It should be noted that in spite of the abundance and wide spread use of Si, it is difficult to get Si powder suitable for plasma spraying. Hence, we have experimented with silicon powder from different sources. (Alfa Chemicals, Germany, Wacker, Germany,

Fig.1. A simple experimental set-up used in the author's laboratory for plasma spray deposition. The water cooled shutter is not shown.

Cerac, U.S. A., etc). Recently we have used Si powder (particle size 10 to 50 microns) supplied by Ethyl Corporation, U. S. A. The particles had irregular shapes. Neutron activation was carried out to determine the impurity levels. In order to obtain deposits of p-type conductivity , calculated amount of B_2O_3 powder was thoroughly mixed with the starting Si powder and in order to obtain n-type conductivity phosphorous powder was added (2). In addition to alumina, we have used Si single crystal, Si polycrystal(SILSO from Wacker, Germany) as substrates.

For preparing the superconducting material YBaCuO, we used the powder supplied by METCO. The particle size varied from 5 to 80 microns. Most of the particles had regular shapes (spherical) but each particle was a composite (3) made up of several smaller particles .

3. Results and discussion

3.1 Silicon deposition

Various substrates such as alumina, copper and stainless steel are placed at 10 to 15 cm in front of the torch and the deposits have been obtained for times varying between 15 to 120 sec. A combination of different parameters such as composition and rate of flow of plasma gas, arc current, powder particle size and rate of powder feed determine the quality of the deposit produced. The influence of some of these parameters on the resistivity p of silicon layers has been reported (2). In the present set-up, the deposits were obtained under the following conditions:

Plasma torch - substrate distance	:	12 cm
Electric arc power	:	380 A, 55 V
Hydrogen flow rate	:	0.3 m^3h^{-1}.
Argon flow rate	:	2.7 m^3h^{-1}.
Substrate temperature	:	600° C .
Substrate preheating	:	3 minutes
Powder feed rate	:	8 g/min

Usually one could detach the deposits from the alumina substrates thus producing free standing samples. Recently we have obtained 50*50*0.2mm silicon samples (4).

The capacitance C of an Si-Hg junction formed on the silicon layers was measured as a function of polarization (V) for several levels of P doping. As an example, figure 2 shows$1/C^2$ as a function of polarization for three doping conditions. A linear variation in $1/C^2$ as a function of V allowed us to calculate the doping concentration N_d. Figure 3 shows the real P concentration c in the Si powder (the quantity added to the Si powder) as a function of N_d. Figure 4 shows $\log(c/N_d)$ as a function of log%100P(or B) in Si.The layers without any addition of P showed p type conductivity. About 0.1 wt% phosphorous in silicon was required to change the conductivity to n-type. The fact that N_d was always smaller than c could be understood due to the fact that some of P burnt away in the plasma flame. Further, initially for small quantities P is in grain boundaries. As the quantity of P increases, mor of it becomes

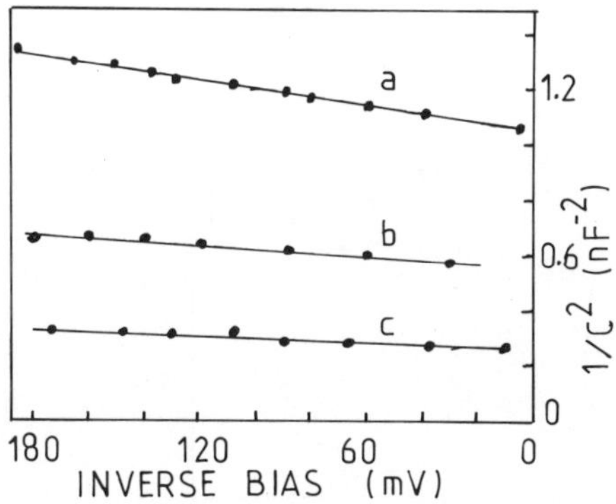

Fig. 2. Plot of $1/C^2$ as a function of the reverse bias voltage of poly Si-Hg diodes for several phosphorous concentrations: (a) 0.5 Wt%, N_d=8. 10^{18} cm^{-3}; (b) 1 Wt%, N_d=1.10^{19} cm^{-3}; (c) 3 Wt%, N_d =5.10^{19} cm^{-3} (ref. 5).

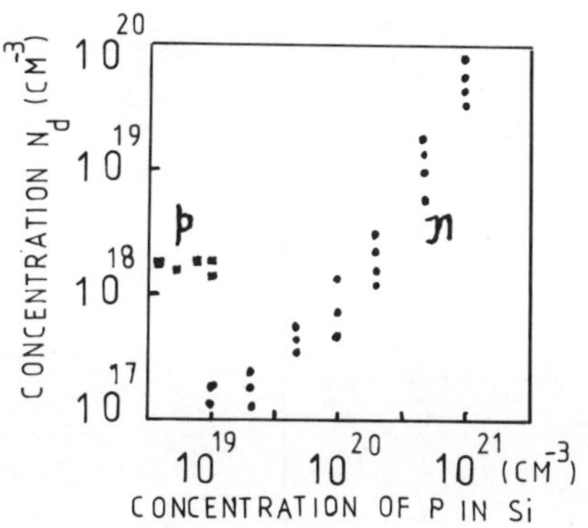

Fig. 3. Dopant concentration N_d as a function of the actual P content in Si powder. Absicssa should be multiplied by 4.5 (ref. 5).

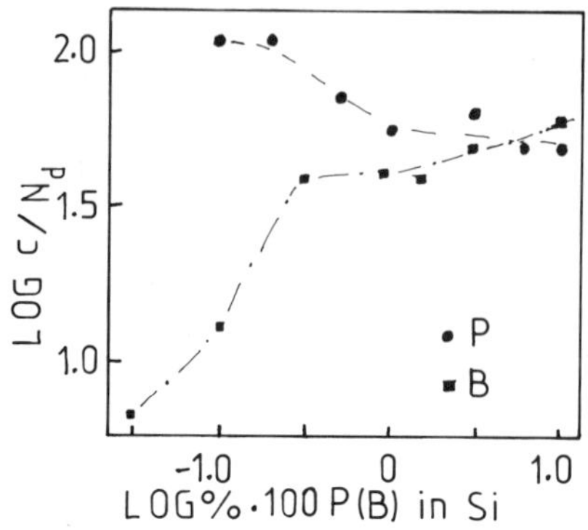

Fig. 4. Plot of log (c/N_d) as a function of P or B concentration expressed as log(100c) where c is the concentration of P or B in Si in weight percent (ref. 5).

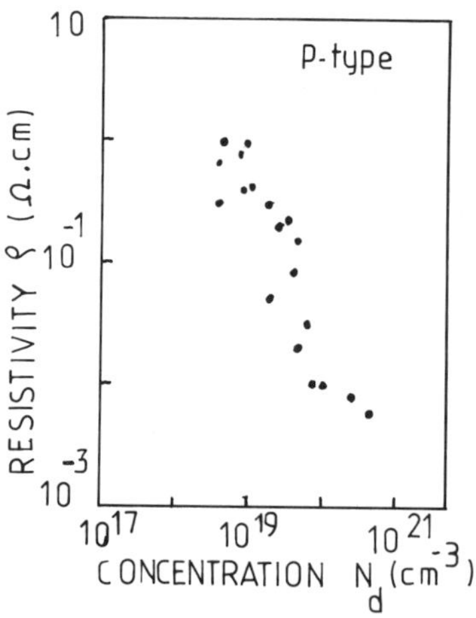

Fig. 5. Resistivity of p-type poly Si layers as a function of N_d (ref.5).

available in the grain and thus the resistivity (ρ) of the layer changes. We could bring about changes of 5 orders of magnitude of in the resistivity. As an example the variation in resistivity (ρ) as a function of B concentration N_d determined by capacitance techniques is shown in figure 5. The mobility of these layers was less than 50 $cm^2V^{-1}s^{-1}$ and hardly varied as a function of the doping concentration. The observed microstructure and the presence of impurities are responsible for the low mobility. The as sprayed layers contained pores and microcracks and the grain size never exceeded 50 microns. The results of neutron activation analyses are given in table 1 at various stages of spraying.

Table 1. Impurity content in ppm in Si at different stages of spraying

Element	Starting powder	After passing through the distributor	After passing through the plasma flame
Cu	0.11	3.2	9.8
Fe	2.2	8	31
Ti	<0.3	5	21
Cr	0.8	1.5	5.7
W	<0.001	<0.002	0.013
Zn	<0.006	1.4	0.8

It is clear from the above table that the distributor and the other parts of the torch increase the impurities considerably which explains the low mobility of the deposits. However, several Schottky diodes were fabricated on these deposits in order to measure the minority carrier diffusion length (L_n) by laser beam induced current (LBIC) technique. After careful cleaning, the silicon surface was thermally oxidized (400°C, 20 min in dry oxygen). Several aluminium dots of area 4 mm^2 and 1000 Å thick were evaporated on to these deposits in ultra high vacuum conditions. A minute copper contact was attached by silver paste to the aluminium dot while another Ga-Al contact was soldered to bare silicon. The photo electric current generated as a result of He-Ne laser irradiation was

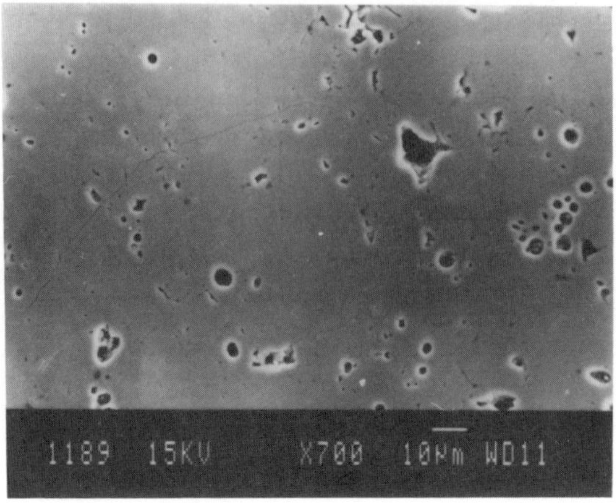

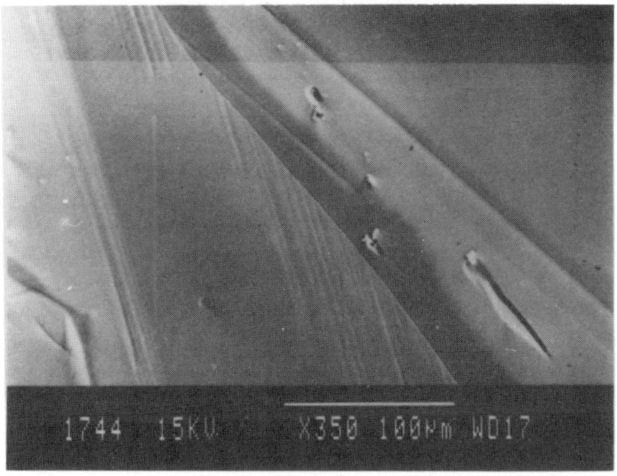

Fig. 6. (above) SEM photograph of as sparayed Si layer. (below) The same layer after electrom beam glazing.

measured as a function of distance d from the edge of aluminium dot. The data were analysed and we found L_n varying between 10 and 15 microns (6).

Two methods were considered to increase the grain size of the deposits. The first was by electron beam glazing (7) and the second was by simply heating in air using halogen lamps (8). The first method uses an electron beam (3-20 keV, 10-30 mA) in high vacuum (10^{-7} Torr) to establish a stable molten zone which induces three main refining processes: evaporation in vacuum, zone melting and slagging. The deposits were subjected to scanning of the pulsed electron beam. The grain size of the deposit increased remarkably after this process (fig. 6). Grain dimensions as large as 1 mm and above in length and 100 microns in width were obtained (9). The resistivity decreased from 1 ohm cm to 0.02 ohm cm and the mobility increased to 320 $cm^2V^{-1}s^{-1}$. Further, the deposits were quite stable mechanically. The LBIC measurements on the recrystallized deposits indicated a high value of L_n ranging from 30 to 66 microns(4). However, on a batch process , these values were not obtained. Usually, L_n was about 30 microns but the deposits seemed to have a uniform doping .

The second method consisted of recrystallizing the as sprayed deposits by zone melting using halogen lamps (2.5 kW). The lamps are slowly displaced along the deposit whose surface area was 25 cm^2. The deposit was mechanically stable after recrystallization and the grain size could be increased to over 1 cm in length and 500 microns across (fig. 7). The sample was cut into 3 or four pieces Hall and resistivity measurements were done on them. A uniform value of hole concentration (10^{17} cm^{-3}) was found on all these pieces and the room temperature mobility was around 225 $cm^2V^{-1}s^{-1}$. The minority carrier diffusion length L_n was around 24 microns(10). The relatively low value of L_n should be related to the presence of metallic impurities. Further work is necessary in order to reduce these impurities which are brought in by the equipment. The powder distributor and the tube connecting the distributor and the torch should be modified and more care should be taken in plasma spraying process. It is interesting to note, however, encouraging results have been obtained without using any argon protection

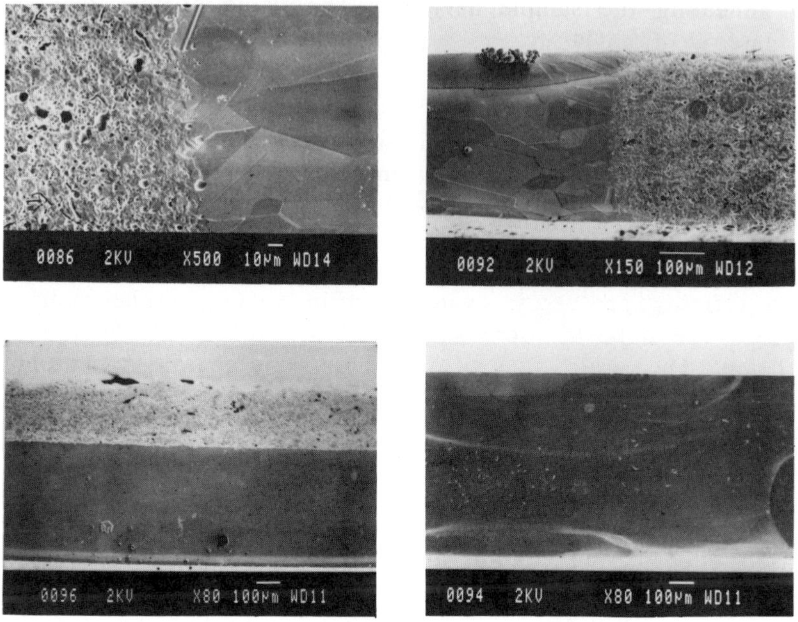

Fig. 7. Self supported Si layer recrystallized by lamps. (above left) surface: (above right) cross section. Si layer deposited on (111) Si single crystal substrate before (bottom left) and after (bottom right) recrystallization.

during spraying. This certainly raises the hope of building solar cells and other devices from electron beam glazed plasma sprayed silicon.

3.2 Crystalline magnetic materials

Polycrystalline ferrite layers could be used in applications such as microwave amplifiers, filters, phase shifters etc. We would like to mention the results obtained by other authors in this field (11,12,13). The influence of process parameters on the magnetic properties of Mn-Mg and Mn-Zn ferrites was studied in some detail by Maeda et al. (13). For example, the saturation magnetic moment of the as sprayed sample increases from a low value of 10 to 36 emu/g by annealing the sample at 800C. The value of the coercive force decreases to 20 Oe as the annealing temperature is increased to 800C. However, if the substrate was heated to 1000° C during the deposition, a smooth surface develops and the coercive force further decreases to 1.5 Oe which is quite close to the bulk value of 0.68 Oe. These properties could be improved considerably by recrystallizing these layers by halogen lamps as was shown above in the case of Si layers.

3.3 Sm-Co permanent magnet

Sm-Co permanent magnets are widely used in different applications. The formation and the stability of the phase $SmCo_5$ phase controls the magnetic properties and *a priori* one would think plasma spray is not well suited to make such a material. However, K.Kumar et al. (14) have obtained high quality magnets by plasma spraying the powders in a protective environment. A close correlation was found between the observed microstructure and the crystalline structure of the deposits. A careful heat treatment at temperatures less than 950°C produces fine grained microstructure which gives a higher value of energy product (BH_{max}) approaching 7 mGOe. Further details can be found elsewhere in this volume. It will be interesting to develop other magnets such as the recently discovered NdFeB by plasma spraying.

3.4 Amorphous magnetic materials

There has not been much work devoted to this area though it is not difficult to imagine that the quenching produced in the plasma spraying could yield amorphous materials. The earliest attempt was on a non-magnetic Zr-Cu amorphous alloy sheet (15). The FeNiPB system was studied by gas melt spraying method (16). Further work was reported using plasma spray by Miura et al (17). Though these authors showed the possibility of quenching using the spray technique by X ray and thermal analyses, no magnetic data were reported. Mention should also be made on the work reported on NiMoCr alloy (18). It should be added that these materials having a proper Cr content also have interesting corrosion properties. Further, by carefully adjusting the ratio of Mo/Cr the wear properties also seem to improve. In our laboratory, preliminary experiments show that the temperature of the substrate and the thickness of the deposit influence the crystallinity of the deposits. The author feels that this area has not been explored enough and further work will be welcome.

3.5 Superconductors

One of the earliest attempts to obtain a superconducting material was reported by Moss et al (19) who prepared an alloy of 60% Mo 40% Ru with a transition temperature T_c of 8.7K. The applications were not forthcoming mainly because of very low temperatures involved. However, the recent discovery of superconductivity above the liquid nitrogen temperature in $YBa_2Cu_3O_{7-x}$ (denoted as YBaCuO) has rekindled the hopes of widespread applications . Further, the material being an oxide, plasma spray is well suited to prepare it on large surfaces without any protective atmosphere. The applications include magnetic shielding, energy storage, bolometers, microwave cavities, sputter deposition sources etc. Several laboratories including ours have since obtained superconducting YBaCuO deposits by plasma spray (20 - 23). Two detailed reports by Swaminathan et al and Danroc et al are presented in this volume. Results obtained on YBaCuO and BiSrCaCuO in our laboratory and elsewhere are briefly presented here.

The YBaCuO samples deposited on alumina substrates in our laboratory could be easily removed to obtain free standing samples. The as sprayed samples were not well crystallised. On the other hand, the annealed samples (1000° C in air followed by 450° C in oxygen for 72 h) were well crystallised and the X ray analyses gave lattice parameters as a=3.83, b=3.886 and c=11.678 Å. A sharp transition was observed(sample #2, fig.8) and the zero resistivity was observed at 92.5 K with a width of only 1.2 K. However, when the heat treatment was carried out a lower temperature at 900° C followed by similar oxygen annealing, the zero resistivity was observed at 87 K with a width of 2 K (sample #1). The field cooled (FC) and zero field cooled (ZFC) susceptibility of #2 measured in a field of 10 Oe is shown in fig.9. The diamagnetic onset is around 90 K. In the FC run, the applied field of 10 Oe is expelled for $T < 80$ K. The ZFC data show that the flux starts penetrating for $T > 40$ K indicating a rather weak coupling between the grains. This would reduce the intergranular critical current density. Assuming a demagnetisation factor of 1/3 , the Meissner effect can be estimated

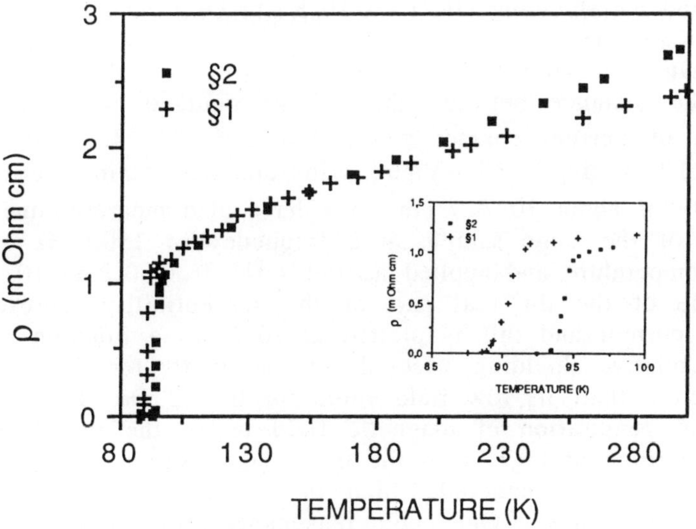

Fig. 8. Resistivity of heat treated plasma sprayed YBaCuO samples (1 and 2) as a function of temperature; expanded view in inset.

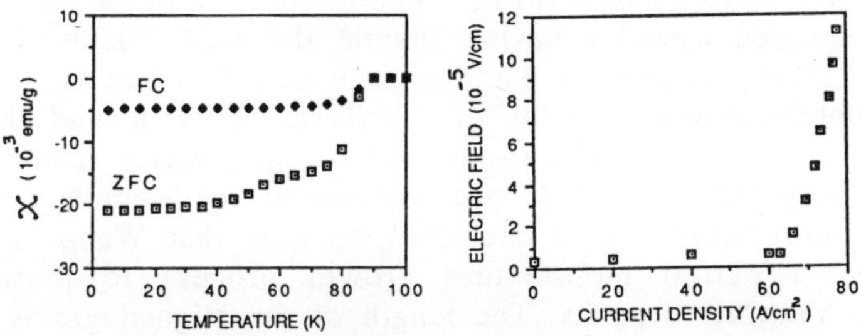

Fig. 9. (left) Field cooled(FC) and Zero field cooled (ZFC) susceptibility of YBaCuO measured in H= 5 Oe.
Fig. 10. (right) Current density of the sample 2 as a function of electric field at 77 K.

to be 10 % and the shielding effect (at 10 K) to be around 50 %. The transport critical current density of the same sample was measured by R. Kormann of Thomson CSF, Orsay using a pulsed current technique. The distance between the voltage contacts was 0.5 cm and the area of current contact was 0.036 cm^2. The I-V curve is shown in fig.10. Taking a 10 μV/cm, criterion, we obtain a critical current density of about 70 A / cm^2. We have also measured the ac susceptibility of the same sample at a frequency of 1500 Hz as a function of temperature and applied ac fields (h) from 0.1 to 10 Oe. The amplitude of the the real part of the susceptibility represents the shielding current and this is plotted at 70 K as a function of h (fig.11). The relative shielding effect decreases rapidly for h> 0.3 Oe and is reduced to half its low field value for h = 3 Oe. This again indicates rapid penetration of magnetic field inside the sample due to poor intergrain coupling. From the dc magnetisation data (ZFC) discussed above it was concluded that the superconducting volume fraction was around 50% . One could reasonably estimate that at for fields less than 0.1 Oe, the shielding is complete. This is not very discouraging given the complexity of the plasma sprayed process. R. Müller et al (24) have studied the shielding properties of a YBaCuO cylinder (obtained by plasma spray) 10 cm in length, 11.5 cm diameter and 0.11 cm wall thickness. They showed that such cylinders have better shielding properties compared to conventionally sintered cylinder having double the wall thickness in fields less than 2 Oe and pointed out that such cylinders can be used to shield completely magnetic fields of 1 mTesla at 77 K and thus can be used in SQUID applications. These properties can be improved to a very large extent if one can remelt the sample using halogen lamps, for example. It is interesting to note that Wang et al (25) have used a partial melt/texture growth process to plasma sprayed (in air) YBaCuO deposits. The length of the aligned grains is in the range of 300 to 500 μm. The transport critical current density of such a sample reached 5000 A/cm^2 under an applied field of 1 tesla at 77 K. This certainly raises hopes for large scale applications though a few technical problems have to be solved.

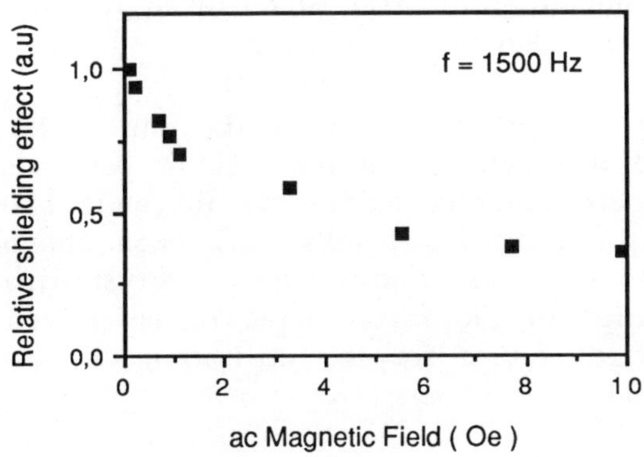

Fig. 11. Relative shielding effect of a heat treated plasma sprayed YBaCuO disk at 70 K as a function of applied ac magnetic field at a fixed frequency of 1500 Hz.

Another superconductor that has been prepared by plasma spraying is the BiSrCaCuO family. The superconducting temperature of $Bi_{1.6}Pb_{0.4}Sr_2Ca_2Cu_3O_y$ could be as high as 110 K, thus 18° higher than that of YBaCuO. Further, it seems that the Bi superconductor does not require tedious annealing procedures in oxygen unlike YBaCuO. However, it seems that it is difficult to stabilise the high T_c phase containing Pb. Often we observe a two step transition indicating the presence of the low T_c(85 K) $Bi_2Sr_2CaCu_2O_y$ phase. But it is encouraging to note that some laboratories (26) by properly processing the starting powder have succeeded in stabilising the high T_c phase obtaining a transition at 103 K. Further work on magnetic properties of such samples should be carried out to assess applications in comparison to YBaCuO.

4. CONCLUSIONS

Whereas plasma spray coatings are being extensively used in the chemical, metallurgical and mechanical engineering areas, the applications involving modern electronic materials are not well developed. This short review indicates three possible areas of application covering materials of widely different properties. These are semiconducting silicon for solar cells and other devices, magnetic oxides and intermetallics in the area of permanent magnet applications and superconducting oxides for magnetic shielding. In all these three areas encouraging results have been obtained. Some technical problems such as reproducible recrystallization and reduction of impurities in the layers have to be solved.Thus the author forsees a bright future for the utilisation of plasma spray technique in this field.

ACKNOWLEDGMENTS

The author thanks with immense pleasure G.BRUN, M.AKANI, B.KAYALI, LE QUANG NAM, J.E.BOUREE, P.PINARD, R.GAUTHIER, M.RODOT, A. EYER, S. SIVOTHTHAMAN and O.GOROCHOV for technical help and discussions. The author is grateful to METCO, France and New York for having kindly supplied the YBaCuO powder. This work was partially funded by contracts EN 3500075 from the European Communities and 8780 N 2036 from CNRS/PIRSEM.

REFERENCES

1. G. Brun, S. Sivoththaman, R. Suryanarayanan and M. Rodot Meas. Sci. Technol. 2 (1991) 985 .

2. R.Suryanarayanan, G.Brun and M.Akani Thin Solid Films, 119 (1984) 67; M.Akani, R.Suryanarayanan and G.Brun J. Appl. Phys., 60 (1986) 457 ;

3. A. M. Bouanani, R. Suryanarayanan and O. Gorochov in Proc. ICMAS 89 (Ed. R. Suryanarayanan, I. I. T. T., Gournay sur Marne, 93460, France, 1989) p283.

4. B. Kayali, M. Rodot, R. Suryanarayanan, Le Quang Nam, R. M'Ghaieth, R. Gauthier and P. Pinard Mat. Sci. Eng. B5 (1991) 51 .

5. M. Akani, R.Suryanarayannan and G.Brun (1987a). Thin Solid Films, 151 (1987) 343 .

6. M.Akani, J.E.Bourée, R.Suryanarayanaan, M.Rodot and G.Brun Solar Cells, 22 (1987) 97 .

7. D. Chambonnet, R. Gauthier, P. Pinard, G. Bechaz and M. Berducat in Poly, Micro and Amorphous Semiconductors, MRS Proceedings (Ed. P. Pinard, Les Editions de Physique, Les Ulis 91 France, 1984) p235.
8. A. Eyer, N. Schillinger, I. Reis and A. Räuber J. Crystal Growth 104 (1990) 119 .
9. R. Suryanarayanan, M. Akani, R. Gauthier, P. Pinard and R. M'Ghaieth Appl. Phys. Lett. 51 (1987) 259 .
10. S. Sivoththaman, M. Rodot, R. Suryanarayanan and A. Eyer Mat. Res. Bull. 27 (1992) 425 .
11. D.H.Harris, R.J.Janowiecki, C.E.Semler, M.C. Willson and J.T. Cheng J. Appl. Phys., 41 (1970) 1348 .
12. R.W.Babbit IEEE Trans. Mag. 11 (1975)11253 .
13. M.H.Maeda, M.Hijikata, S.Kitahara and I.Moromoto Trans. National Res. Inst. Metals, 19 (1977)17 .
14. K.M.Kumar, D.Das and E.Wettstein J. Appl. Phys., 49 (1978) 2052
15. B.G.Giessen, N.M.Madhava, R.J.Murphy, R.Ray and J.Surette Met. Trans. 8A, (1977) 364
16. P.M.Shingu, K.Shimomura and R.Ozaki . Trans. Jap. Inst. Metals, 20, (1979) 33.
17. H.Miura, S.Isa and K.Omuro In Jap. Inst. Metals. Sendai Conf. 1982(Ed. T. Masumoto) p. 43
18. S.K. Das, E.M. Norin and R.L.Bye: in Rapidly Soliddified Metastable Materials (Ed. B.H.Kear , North Holland, Amsterdam,1984)
19. M.Moss., D.L.Smith and R.A. Lefever . Appl. Phys. Lett., 5 (1964)120.
20. J.J.Cuomo, C.R.Guarnnieri, S.A.Sivashankar, R.A. Roy, D.S. Roy, D.S. Yee and R. Rosenberg Adv. Ceram. Mat., 2 (1987) 242.
21.R.Suryanarayanan et al. Le Vide, supplément n° 241 (1988) p21
22. L.S.. Zen, S.W.Qian, Q.Y.Hy, B.H.Yu, H.Z.Zhao, K.Guan, L.S. Fu and Q.Q.Yang . Thin Solid Films 152 (1987) L-143.
23. J.Karthikeyan et al.J. Phys. D Appl. Phys., 21 (1988)1246.
24. R. Müller et al. in Int. Symp. on Superconductivity, Georgenthal, 1990(World Scientific, Singapore, 1990): S. Rohr et al. Modern Physics Lett. B4, (1990) 717.
25. H. Wang et al. Appl. Phys. Lett. 57 (1990) 2495 .
26. E. Lugsheider, T. F. Weber and P. Jokiel in ICMAS 90 (Ed. R. Tournier and R. Suryanarayanan, I.I.T.T, Gournay sur Marne, 93460, France, 1990) p 105.

Chapter 10

STRUCTURE AND PROPERTIES OF SPRAYED SAMARIUM-COBALT MATERIALS

Kaplesh Kumar
The Charles Stark Draper Laboratory, Inc., Cambridge, MA 02139

ABSTRACT

The structure and properties of sprayed and heat treated Sm-Co materials, particularly as they pertain to the fabrication of $SmCo_5$ permanent magnets, are reviewed. Significant results have included the fabrication of the highest quality isotropic magnets and the finding that the as-sprayed amorphous materials are able to absorb more hydrogen, with little physical degradation, compared to their crystalline counterparts. Success at producing anisotropic magnets was only partly achieved, mainly because of difficulties in achieving high alignment and high coercivity simultaneously in the deposits. With additional work, high quality aligned $SmCo_5$ magnets with considerable application potential appear possible. Potential applications of the sprayed Sm-Co materials are briefly discussed.

1. INTRODUCTION

$SmCo_5$ was among the early magnetic and electronic materials to which plasma spraying was applied as a fabrication technology.(1) Previous uses of plasma spraying had centered on its use as a technology for the deposition of hard ceramic type coatings and for the repair of expensive mechanically worn parts.(2,3) Among others, plasma spraying has recently been used for fabricating polycrystalline silicon and liquid nitrogen high temperature superconductor deposits.[4,5]

Willson and Janowiecki attempted to exploit the process-related advantages of this technology in the synthesis of permanent magnets and permanent magnet structures based on $SmCo_5$.[1] The process seemed particularly attractive because of its potential cost-effectiveness. These early efforts, which were mostly of a process development nature, demonstrated that it was feasible to produce $SmCo_5$ magnets, with reasonable properties, in suitable device configurations on selected substrates.

In subsequent work, Kumar and co-workers studied the effects of processing on deposit structure with the aim of optimizing the material behavior.[6,7] Changes in microstructure brought about by the deposition and subsequent heat treatments were used to explain changes in material behavior. These efforts led to the fabrication of the highest quality isotropic $SmCo_5$ magnets produced by any technology. Indeed, the 70 kOe room temperature intrinsic coercivity (H_{ci}) value measured for several of their magnets is also the highest value ever reported for any permanent magnet.[6] Because the magnets produced were isotropic, their energy products (BH_{Max}) were limited to 7-8 MGOe, a little over a third of the values typically seen for conventionally sintered $SmCo_5$ magnets.

Efforts to produce high quality aligned magnets by spraying were only partly successful. A very high degree of alignment was obtained in several of the deposits; however, H_{ci} was seen to decrease concurrently. Continued efforts also yielded the unexpected result that $SmCo_5$, and by analogy other relevant compositions, were capable of storing substantially greater amounts of hydrogen when they were in the amorphous state compared to when they were crystalline.[8,9] Similar results, with a strong potential for application to the "environment-friendly" nickel hydride batteries, have recently appeared in the literature.[10,11] Nickel hydride batteries are currently being marketed as replacements for Ni-Cd batteries, which are used extensively in consumer electronics.

The discussion below reviews the structure and properties of arc plasma sprayed Sm-Co materials in both the as-sprayed condition as well as after post-deposition thermal treatments. Potential application areas and possibilities for additional developmental efforts are briefly discussed towards the end. For additional comprehensive discussions of permanent magnets and magnetics fundamentals, the reader is referred elsewhere.[12-14]

2. MAGNETIC BEHAVIOR

2.1 Magnetism

The unpaired electrons in their partly filled electronic shells, *4f* for Sm and *3d* for Co, are responsible for the observed effects. The motivation for examining these types of compositions resulted from the high magnetic moment seen for the rare earths (such as Sm) at low temperatures and the high Curie temperature ferromagnetic behavior characteristic of the transition metals (such as Co).[15] The net magnetic moments of Sm and Co couple ferromagnetically to each other. The Sm sublattice anisotropy contributes substantially to the extremely large 400 kOe uniaxial magnetocrystalline anisotropy field of the hexagonal $SmCo_5$ crystal along its unique c-axis.[16] The Co sublattice dominates at the high temperatures and contributes to the very high Curie point (725°C) of $SmCo_5$. The Curie temperature of Sm_2Co_{17} is higher still, 920°C.[17]

2.2 Demagnetization Characteristics

The principal parameters used for characterizing permanent magnets using cgs units are shown in Figure 1. The magnetization ($4\pi M$), and the induction ($B= H+4\pi M$) are plotted in Gauss (G) on the vertical axis while the internal magnetic field (H, which represents the vectorial sum of the applied field and the self-demagnetizing field) is plotted in Oersted (Oe) on the horizontal axis. The second quadrant is the most important since it identifies the performance-related attributes of the magnet, after is has been magnetized to saturation along a certain direction by the application of a large enough magnetic field in that direction. The $4\pi M$ curve describes the intrinsic demagnetization behavior of the material. It intersects the vertical axis at its remanence ($4\pi M_R$) and the horizontal axis at H_{ci}, which is a measure of the magnet's resistance to demagnetization. Similarly, the induction curve, which determines the field sensed external to the magnet, intersects the vertical axis at B_R (= $4\pi M_R$) and the field axis at H_c, the inductive coercivity; H_c is typically less than (or at best equal to) H_{ci}. H_K is a measure of the

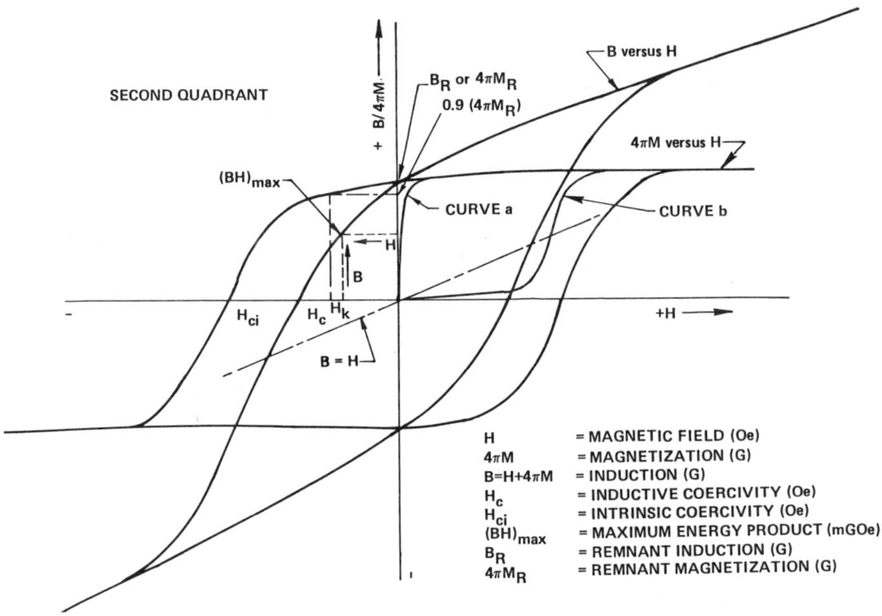

Figure 1. General Hysteresis Curves for Permanent Magnets[12]

squareness of the second quadrant loop. The energy product (BH_{Max}) in MGOe is widely used as a figure of merit; it is simply the maximum product of B and H in the second quadrant.

3. FABRICATION AND STRUCTURE CONSIDERATIONS

3.1 Fabrication

Because the compound Sm_2O_3 has an extremely large negative free energy of formation, it is necessary to spray deposit the Sm-Co alloys in an inert environment. Oxidation leads to localized depletion of Sm and results in poor demagnetization behavior because of the associated drastic reduction of H_{ci}. Inert gases (argon, helium) are preferred for the plasma and the carrier gases to avoid any chemical interaction with the alloy being deposited. For instance, hydrogen is readily absorbed in large quantities at near-ambient pressures and temperatures by $SmCo_5$; the volume efficiency of hydrogen storage is larger in $SmCo_5$ (and even more in $LaNi_5$) than that obtained in the liquid state at cryo temperatures.(18) Because hydrides are formed readily, the crystalline compound develops large stresses which cannot be accommodated and this leads to severe fracturing of the material. Decrepitation techniques have been developed for producing the alloy powder using hydrogen absorption-desorption cycling techniques.(19)

$SmCo_5$ is formed from the liquid state by a peritectic reaction in the range of 1293-1320°C. Fortunately the temperature difference between the liquidus and the peritectic reaction is quite small; it is, thus, possible to cool the alloy through this range without encountering significant chemical segregation through coring effects. The fast quenching rates approaching 10^7°C/sec associated with plasma spraying(20) result in even less segregation and the deposited alloy is readily homogenized at 1000-1050°C temperatures, considerably lower than the >1120°C needed for sintering powdered compacts. Because of vapor pressure differences of the elements, the alloy experiences a several percent loss of Sm; the composition of the starting alloy powder is adjusted to correct for this loss. The sprayed materials of Kumar et al were fabricated with a Metco 3MB plasma spray gun, using either single or blended alloy powders in a narrow particle size distribution range. Helium and argon were used in combination for forming the plasma; argon served as the carrier gas.

3.2 Structure

$SmCo_5$ magnets are conventionally produced using powder metallurgy sintering techniques. The technique involves comminuting the alloy to particle sizes in the range of 1 to 5 μm, aligning and cold-compacting the powder in a 20 kOe magnetic field, and sintering the compacts at about 1120°C in an inert atmosphere. Because of the need to produce fine powders (resulting in individual single crystals) to get good particulate alignment along the c-axis, the oxygen in these magnets increases from a starting alloy content of a few 100s of ppm to typically 0.5-1 wt%. Also, because of the need to densify the compacts (at 1120°C), the resulting sintered magnet grain size is 20-30 μm. (To avoid excessive grain growth, the sinter temperature is controlled to within about 5°C and this results in an as-sintered density of about 92-94% of the theoretical 8.6 g/cm^3.) Sintered magnets also show considerable microcracking. The composition of these magnets needs to be controlled within a very narrow range for developing high H_{ci}; consequently, control is achieved by blending powders of two different compositions.(21) This results in the presence of a number of secondary phases in the microstructure.

As previously noted, oxygen increases lead to easier magnetization reversal and low H_{ci}. Microcracking and grain growth also result in the lowering of H_{ci}. Compositional inhomogeneity makes the magnet susceptible to severe H_{ci} degradation when exposed to intermediate (750°C) temperatures because of which $SmCo_5$ magnets, produced by powder metallurgy, have to be quenched from their post-sinter optimization heat treatment at 900°C.(22) This leads to severe stressing of the material and results in magnets that are less stable over time, compared to those that are capable of being cooled slowly from high temperature.

The optimum microstructure, therefore, consists of one that shows high density, fine well-aligned grains, no microcracks, chemical uniformity, and very low oxygen pick-up. As noted in the introduction, save for the concurrent achievement of high grain alignment, extremely high quality isotropic $SmCo_5$ magnets have been demonstrated with the arc plasma spray technique.

4. STRUCTURE

4.1 Microstructure

The as-sprayed microstructure of an early $SmCo_5$ deposit, formed on a water-cooled rotating copper substrate inside an environmentally controlled chamber, is given in Figure 2. The oxygen content of these deposits was typically under 0.2 wt%, considerably lower than that of the sintered materials. The multiphase layered morphology, typical of plasma sprayed

Figure 2. As-Sprayed Microstructure of an Early Deposit

coatings, is seen along with unmelted and partly melted particles, in addition to the pores that appear as dark circular features. With appropriate process control, considerably improved uniform microstructures were produced. (See Figure 3.) The improvements resulted from the use of a reasonably narrow powder particle size distribution with optimized plasma spray process parameters. Figure 3 shows the effect of varying one process parameter, the coarseness of the starting powder, on the degree of refinement in the microstructure.

When the sprayed materials are exposed to high temperatures, the layered morphology transforms to an equiaxed grain structure through the formation of recrystallized grains and their subsequent growth. Figure 4 shows the effects of exposing samples from one such sprayed deposit to maximum temperatures of 1000°C and 1100°C.[23] In both cases, a compositionally uniform, essentially single phase microstructure is seen to result. The average grain sizes of about 2-5 μm and 10-20 μm for the treatments at 1000°C and 1100°C, respectively, show that it is possible to limit

(a) (b)

Figure 3. Effect of Starting Powder Particle Size (a) Coarse, (b) Fine

(a) (b)

Figure 4. Grain Size Dependence on Heat Treatment Temperature
(a) 1000°C, (b) 1100°C

the grain size of these deposits by limiting the maximum temperature of exposure. The minimum in exposure temperature was determined by the necessity to achieve maximum properties and these were seen to develop at temperatures of about 1000°C. These microstructures were also free of any evidence of microcracking.

The as-sprayed density of the $SmCo_5$ materials was also quite high; with the wet displacement technique these were measured about 97% of the theoretical. Consistent with this high measured value of the density and the associated expectation of the absence of an interconnected porous structure, sprayed samples were found to densify to nearly full-density upon direct exposure to 15 kpsi argon gas pressure at 900°C in a hot isostatic pressing autoclave.[23]

Apart from these several attributes, the sprayed magnets also showed a remarkable resistance to magnetic property degradation from exposures to intermediate temperatures following the high temperature optimizing treatment (which for these materials is in the 1000°C range). The change in microstructure seen from extended (weeks of) exposure to intermediate temperatures is exemplified in Figure 5. A fine scale distribution of second phase (Sm_2Co_{17}) particles is randomly located in the structure and striations occur in the interior of the grains. The second phase precipitation most likely results from a narrowing chemical homogeneity range for $SmCo_5$, with decreasing temperatures, and the striations are the result of plastic deformation processes. (Sintered magnets show much more extensive second phase precipitation, possibly because of a considerably larger chemical inhomogeneity, along with striation formation.) Researchers had previously

Figure 5. Microstructure After Several Weeks Exposure to 650°C and 700°C

postulated that these metallographic features were the result of a eutectoidal decomposition of the $SmCo_5$ phase.[24] More recent work discounts this possibility; on balance, it appears that $SmCo_5$ may indeed be stable.[12]

4.2 Crystal Structure

Figure 6(a) shows the x-ray diffraction pattern of an as-sprayed deposit formed on the water cooled copper substrate. The side examined was that which was adjacent to the substrate. The patterns obtained from these materials were basically characterized by several very broad unresolved peaks indicative of the materials being in a near-amorphous state. As expected, a greater degree of crystallinity was seen to result in the patterns as the deposit build-up in thickness; however, for the most part, the deposits remained near-amorphous for thicknesses as large as a few millimeters. Figure 6 also shows the x-ray pattern obtained from a crystalline deposit, fabricated on a hotter substrate. When these near-amorphous deposits are heated to the high temperatures, patterns with well-defined peaks corresponding to the equilibrium $SmCo_5$ structure are seen to evolve. The deposits, however, continue to be nearly isotropic through the evolution of their structure.

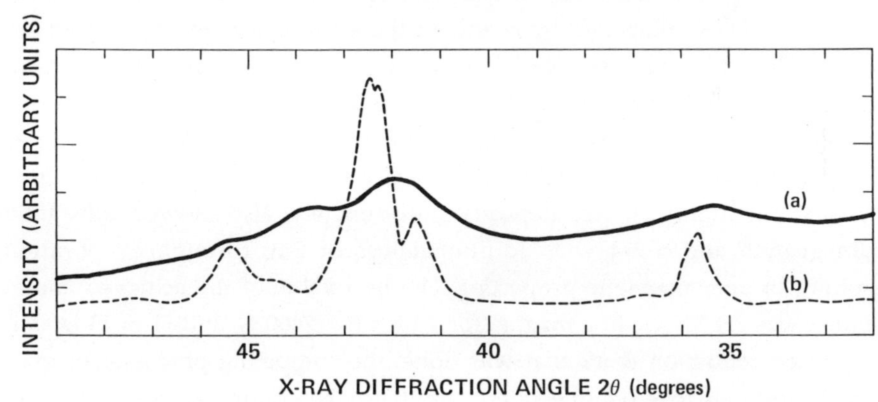

Figure 6. X-ray Diffraction Patterns for (a) Amorphous, and (b) Crystalline Deposits; CuK_α radiation.

Because, as discussed previously, the degree of the crystallographic alignment of the grains determines the magnetization and, consequently, the energy product, a considerable level of effort was expended in attempting to produce aligned sprayed $SmCo_5$ deposits. The early efforts included trying to induce orientation through the control of the crystallization and grain growth processes in the as-sprayed nearly-amorphous 1Sm-5Co materials. The as-sprayed deposits were either heat treated in the presence of an applied 80 kOe static dc magnetic field or subjected to increasing temperatures on one face, while the other was kept chilled to promote recrystallization and grain growth while under a temperature gradient. Neither procedure resulted in significant change in orientation or magnetic behavior.

Success in producing highly oriented deposits was finally achieved when the materials were deposited at very high (> 850°C) substrate temperatures.[25] These materials were in their nearly-equilibrium crystalline state, as-deposited, and showed an increasing degree of alignment with increasing substrate temperature. (See Figure 7.) A strong temperature gradient was imposed on the deposit as it built-up. The depositions were performed on the flat head of steel substrates, shaped in the form of bolts. The substrate temperature and the temperature gradient under which the deposit was formed were controlled by changing the stem length and stem diameter of the bolt, which was screwed into a water-cooled rotating copper head. Nearly perfect alignment was achieved in several of the deposits. The c-axis magnetic domain structure in Figure 8, obtained using the magnetoptic Kerr effect and providing corroborating evidence, was typical of the highly aligned deposits. The highly aligned deposits, unfortunately, also showed substantial grain growth and a 2-4 wt% additional loss of Sm, through evaporation, resulting in poor magnetic properties. The best value of the achieved energy product was 9.4 MGOe in a magnet for which B_r= 7400 G, and H_{ci}= 13 kOe. In the limited follow-on work that was done, the competing processes of grain growth and Sm loss on one hand could not be optimized with the high alignment requirement on the other. It appeared that use of a higher energy

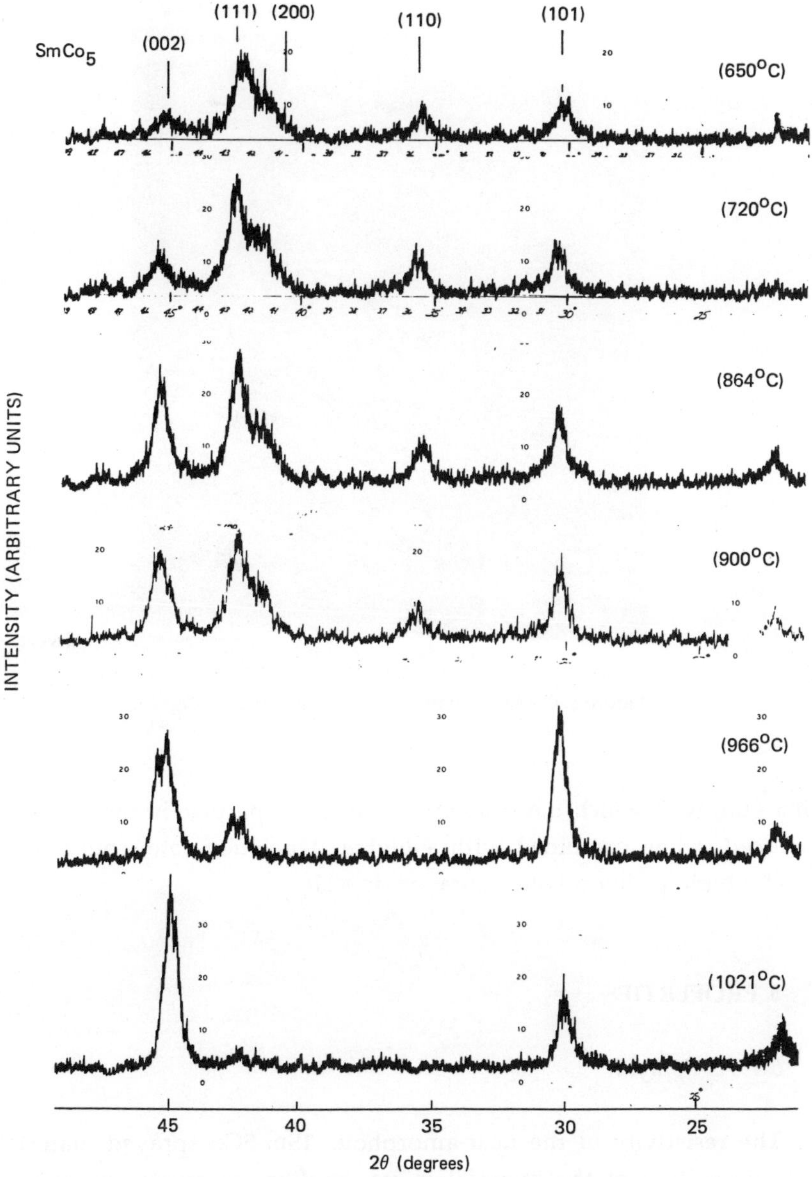

Figure 7. Effect of Substrate Temperature on Alignment[25]

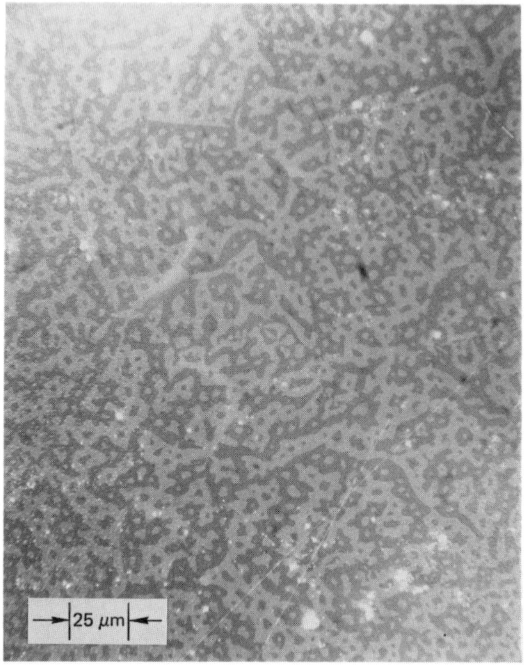

Figure 8. C-axis Domain Pattern on an Aligned Deposit

plasma gun, with which the residence time of the particle in the flame could be reduced, when combined with a higher deposition rate, could possibly result in a high quality oriented sprayed deposit.

5. PROPERTIES

5.1 As-Sprayed

The resistivity of the near-amorphous 1Sm-5Co sprayed materials is several times that of the material in its equilibrium crystalline state. The near-amorphous deposits could be produced in thicknesses of over 0.1 in and unlike the crystalline materials, which are extremely brittle, they could be machined using single point tooling. Their lathe machining characteristics were similar to that of cast iron.[8]

The near-amorphous deposits typically showed a low remanence (4.5-5.0 G) and a substantial H_{ci} (~ 20 kOe) after magnetization with applied fields of 140 kOe. They were also extremely difficult to magnetize to saturation. This behavior is believed to have resulted from the impediment to the motion of the domain wall offered by compositional inhomogeneities and quenched-in internal stresses in the deposits.

When the near-amorphous materials were subjected to thermal treatments in argon at temperatures of less than about 500°C, at which $SmCo_5$ is known to crystallize from its amorphous state, little change was seen in the magnetic behavior. When, instead, the deposits were similarly heat treated in hydrogen, easy approach to saturation coupled with a considerable lowering of H_{ci} was seen to result.[8] This behavior was interpreted to result from a facilitation of the atomic rearrangement process brought about by treating the material in hydrogen. (The results of powder metallurgy sinter studies in hydrogen[26] had similarly implied a substantially higher mobility of the atoms in the presence of hydrogen, compared to argon.) Exposure to hydrogen also resulted in raising the crystallization temperature to about 700°C and showed an increase in the specific heat, C_p, whose value was used to monitor the extent of the interaction of the materials with hydrogen. Unlike their crystalline counterparts, the amorphous deposits were extremely resistant to physical degradation from absorption-desorption of hydrogen. These efforts resulted in the remarkable finding that the alloys in their amorphous state were more suited to hydrogen storage than their crystalline counterparts.[8,9] (Conceivably, the deposits could be produced with hydrogen as a plasma gas eliminating the need for any subsequent surface activation.) As stated previously, reports of similar findings have since appeared in the literature.

5.2 Heat Treated

Several hour exposures to temperatures of 1000°C to 1050°C result in remarkable improvements in magnetic properties. No subsequent treatment at the lower 900°C temperature (such as that required with sintered magnets) is needed for these materials. Excellent second quadrant loops with room temperature H_{ci} values exceeding 50 to 60 kOe were routinely seen for the heat treated materials; many deposits had values of about 70 kOe. These values are the highest reported for any permanent magnet.[6] The very high values are believed to result from the low oxygen content (< 0.2 wt%), fine grain size (2-5 μm), no microcracks, a nearly single phase microstructure, and the isotropic nature of the deposit. (For comparison, the highest H_{ci} reported for sintered $SmCo_5$ magnets is 43 kOe.) Typically, good sprayed and heat treated isotropic $SmCo_5$ magnets have B_r= 5500-5600 G, H_{ci}= 50-70 kOe, and BH_{Max}= 7-8 MGOe. Commercial sintered magnets have B_r= 8000-9000 G, H_{ci}= 20-35 kOe, and BH_{Max}~ 18 MGOe.

Unlike sintered $SmCo_5$ magnets, which show poor second quadrant characteristics unless they are cooled rapidly from 900°C, the sprayed magnets show little to no deterioration when cooled slowly by shutting-off the furnace after their 1000°C thermal optimization treatment. This provides for a material that is minimally stressed, has no microcracks, and demonstrates higher mechanical integrity. The most severe property degradation in sintered $SmCo_5$ magnets occurs at about 750°C where, as discussed previously, extensive second (Sm_2Co_{17}) phase precipitation occurs. Sprayed $SmCo_5$ magnets show a remarkable resistance to property degradation at these temperatures, even for long-duration exposures on the order of a few weeks.[27] Similar to observations made on sintered magnets, a hysteresis is observed in H_{ci} when they are exposed for short time (hours) to decreasing and subsequently increasing temperatures. This hysteretic behavior in H_{ci} is consistent with the known narrowing of the homogeneity range of $SmCo_5$ with decreasing temperature and has been explained as associated with second phase (Sm_2Co_{17}) precipitation and dissolution mechanisms; the hysteresis in H_{ci} results from the difference in the rates of these two metallurgical processes.[28] Consistent with this explanation, little to no hysteresis is seen when the exposure times are long (days to weeks).[27,28]

6. SUMMARY AND POTENTIAL APPLICATIONS

As-sprayed amorphous 1Sm-5Co has been shown to absorb considerably more hydrogen than its crystalline counterpart, with little physical degradation. In the near-term, similarly processed materials (such as amorphous $LaNi_5$) are strong candidates for improved nickel hydride batteries, which are being marketed to replace the widely used Ni-Cd.

High quality isotropic $SmCo_5$ magnets, with record H_{ci} (~70 kOe), low oxygen (< 0.2 wt%), and fine-grained (2-5 μm) nearly single-phase microstructures, are produced by heat treating the deposits at about 1000°C. The magnets can be cooled slowly from the elevated temperatures with few deleterious effects. Near-net shape sprayed and heat treated $SmCo_5$ configurations are strong cost-competitive alternatives in applications employing Pt-Co or plastic-bonded and isotropic Sm-Co magnets.

The limited work done has suggested the feasibility of producing aligned $SmCo_5$ magnets. If successfully developed, such magnets would find extensive usage in numerous applications such as motors, actuators, and electronics. The cost-effectiveness of the spray process is amplified by its ability to fabricate the structures in near-net shape geometries.

7. ACKNOWLEDGEMENTS

The author thanks P. Halloran and J. Crane for their help with the figures and the formatting of this manuscript.

REFERENCES

1. M. C. Willson and R. J. Janowiecki, AIP Conf. Proc. **24** (1975) 689.

2. D. A. Gerdeman and N. L. Hecht, *Arc Plasma Technology in Materials Science* (Springer-Verlag/Wien, New York, 1972).

3. H. Herman, Scientific American **259** (Sept. 1988) 112.

4. R. Suryanarayanan, Proc. Conf. Internationale PLASMA 88, Paris, France, 14-15 Nov. 1988 (Pergamon Press).

5. W. T. Elam, J. P. Kirkland, R. A. Neiser, E. F. Skelton, S. Sampath, and H. Herman, ADVANCED CERAMIC MATERIALS **2** (1987) 411.

6. K. Kumar, D. Das, and E. Wettstein, J. Appl. Phys. **49** (1978) 2052.

7. K. Kumar and D. Das, Thin Solid Films **54** (1978) 263.

8. K. Kumar, D. Das, and R. Williams, J. Appl. Phys. **51** (1980) 1031.

9. K. Kumar, "Method of Storing Hydrogen Using Nonequilibrium Materials and System," U. S. Patent 4,565,686, issued 21 January 1986.

10. K. Ramakrishna, S. K. Singh, A. K. Singh, and O. N. Srivastava, Prog. in Hydrogen Energy: Proc. Natl. Workshop on Hydrogen Energy, New Delhi, India, 4-6 July 1985 (Dordrecht D. Reidel Publishing Co., 1987) p. 81; I. P. Jain, Y. K. Vijay, L. K. Malhotra, and K. S. Uppadhyay, Int. J. Hydrogen Energy **13**, (1988) 15.

11. "Japan NMH Battery Boost," in *Electron. Engg. Times*, (16 March 1992)

12. K. Kumar, J. Appl. Phys. **63** (1988) R13.

13. K. J. Strnat, in *Ferromagnetic Materials Vol. 4* (North-Holland, 1988).

14. J. D, Livingston, Journal of Metals (February 1990) 30.

15. G. Hoffer and K. Strnat, IEEE Trans. Magn. **MAG-2,** (1966) 487.

16. A. V. Deryagin, E. A. Barbanova, and A. I. Ulyanov, Phys. Status Solidi **A 31,** (1975) 391.

17. E. A. Nesbitt and J. H. Wernick, *Rare Earth Permanent Magnets,* (Academic Press, New York, 1973).

18. M. V. C. Sastri and V. V. V. N. S. Ramakrishna Rao, Trans. Indian Inst. Metals **37,** (1984) 273.

19. A. Kianvash and I. R. Harris, J. Mater. Sci. **19** (1984) 353.

20. M. Moss, D. L. Smith, and R. A. Lefever, Appl. Phys. Lett. **5,** (1964).

21. M. G. Benz and D. L. Martin, J. Appl. Phys. **43** (1972) 3165.

22. F. J. A. den Broeder and H. Zijlstra, J. Appl. Phys. **47** (1976) 2688.

23. D. Das and K. Kumar, *Proc. 4th Intl Workshop on Rare Earth-Cobalt Magnets and Their Applications,* edited by K. Strnat (1978) p. 494.

24. F. J. A. den Broeder and K. H. J. Buschow, J. Less-Common Metals **29** (1972) 65.

25. K. Kumar and D. Das, J. Appl. Phys. **60,** (1986) 3779.

26. F. G. Jones, M. Doser, and T. Nezu, IEEE Trans. Magn. **MAG-13** (1977) 1320.

27. K. Kumar, D. Das, and E. Wettstein, IEEE Trans. Magn. **MAG-14** (1978) 788.

28. K. Kumar and D. Das, J. Appl. Phys. **50,** (1979) 2940.

Chapter 11

Preparation of $YBa_2Cu_3O_7$ High Tc Superconducting Coatings by Plasma Spraying

J. Danroc and J. Lacombe
CEA- CEREM
CENG 85 X
38041 GRENOBLE CEDEX

1. INTRODUCTION

Following the discovery by J. G. Bendorz and K. A. Muller (1) of high critical temperature superconductivity in perovskite oxides, a large number of both theoretical and experimental research programmes were set up.

In 1987, Wu et al. (2), discovered that the compound $YBa_2Cu_3O_7$ showed superconductivity above 90 K; a new era has therefore been reached, where liquid nitrogen (T = 77 K) can be used as coolant instead of much more costly liquid helium (T = 4.2 K). This reduction in the cost and ease of use of liquid nitrogen opens the way to the rapid development of these materials.

Meanwhile, the development of these materials is linked to the electrical characteristics of the products made and few results were known at the time. Power applications (loss-free transport of electrical current, production of intense magnetic fields...), call for critical current densities Jc and critical magnetic field densities Hc_2 as high as possible (of the order of 10^5 to 10^6 A/cm^2 for Jc).

Another important factor limiting the development of these ceramics is the difficulty in forming.

For this reason, studies of the preparation of coatings by thermal spraying have been undertaken. For many applications, bulk material can be replaced to advantage by a "thick coating". Thermal spraying, because its flexibility and versatility seems especially well adapted for making coatings for industrial applications.

In this article we describe:
- the preparation of $YBa_2Cu_3O_7$ powders for spraying
- the thermal spraying of these powders using various techniques
- the physicochemical and electrical characterisation of the coatings obtained.

The results concern work published in the literature and more especially the studies undertaken at CEA - CEREM.

2. THE COMPOUND $YBa_2Cu_3O_{7-\delta}$

2.1 Structure

The structure of $YBa_2Cu_3O_{7-\delta}$ can be described as a perovskite lattice triply deficient in oxygen (1, 2). The oxygen stoichiometry in this compound can vary between O_6 and O_7. The structures of the limiting compounds ($\delta = 0$, $\delta = 1$) have orthorhombic (3) and tetragonal (4) symmetries.

Structural transformation

During thermal treatment (e.g. 950°C in air followed by a slow cooling), the compound $YBa_2Cu_3O_{7-\delta}$ in the tetragonal phase at high temperature transforms to the orthorhombic phase at low temperature. This orthorhombic-tetragonal transformation is a function of the temperature and the oxygen partial pressure. It occurs between 580°C (5) 660°C (6) in air. The structural change is accompanied by a variation in the lattice volume of the order of 0.20% which often causes microcracking in samples.

2.2 Critical temperature

The critical temperature of $YBa_2Cu_3O_{7-\delta}$ depends on the oxygen stoïchiometry and homogeneity of the oxygen atom distribution in the structure. The main characteristics of the compounds are as follows:

- Tc = 92 K when $\delta = 0$
- there is a plateau at ~ Tc = 60 K for $0.3 < \delta < 0.5$
- the compound is not superconducting for $\delta > 0.7$

2.3 Critical current density

The critical current density, Jc, of a superconducting sample, which is an intrinsic property of the material is generally linked with its microstructure and therefore with its preparation method. Comparative values of Jc (77 K, H = 0T) of epitaxial thin films, textured samples and non-textured sintered samples are given in table 1. Large differences in critical current density along different crystallographic axes are also shown for oriented samples.

	epitaxial thin films (7)	**textured polycrystalline**	**non-textured polycrystalline**
Jc (A/cm²)	10^6 in the a, b plane	10^6 in the a, b plane	10^3
77 K, 0T	10^4 along c	5×10^4 along c	

Table 1

2.4 *Phase equilibria in the YBaCuO system*

During the process of plasma spraying, the compound $YBa_2Cu_3O_{7-\delta}$ is heated to a high temperature and then quenched (quench rate : 10^6 k/.s). After spraying, the coatings are microcrystalline or amorphous, and must be heat treated to 950°C to restore the superconducting phase. It it therefore necessary to know the reactions and decompositions likely to occur at high temperature and to identify the phases which might appear during the heat treatment. At 950°C in air, $YBa_2Cu_3O_{7-\delta}$ is in equilibrium with the phases Y_2BaCuO_5, $BaCuO_2$ and CuO.

At higher temperatures, the following reactions occur:

– near 1010°C: $YBa_2Cu_3O_{7-\delta}$ tetragonal form ($\delta = 1$) melts incongruently in 0.2-0.3 atmospheres of oxygen (9) :

$$2 < YBa_2Cu_3O_6 > + 1/2\ [O_2] <\!\!-\!\!-\!\!> < Y_2BaCuO_5 > + (3BaO - 5CuO)$$

– near 1240°C: Y_2BaCuO_5 melts at the peritectic and decomposes into Y_2O_3 + liquid (10)

3. PREPARATION OF $YBa_2Cu_3O_7$ COATINGS

In this chapter we study more precisely the steps of the preparation process developed and present the characteristics of the coatings obtained.

3-1 *General organisation of the preparation process*

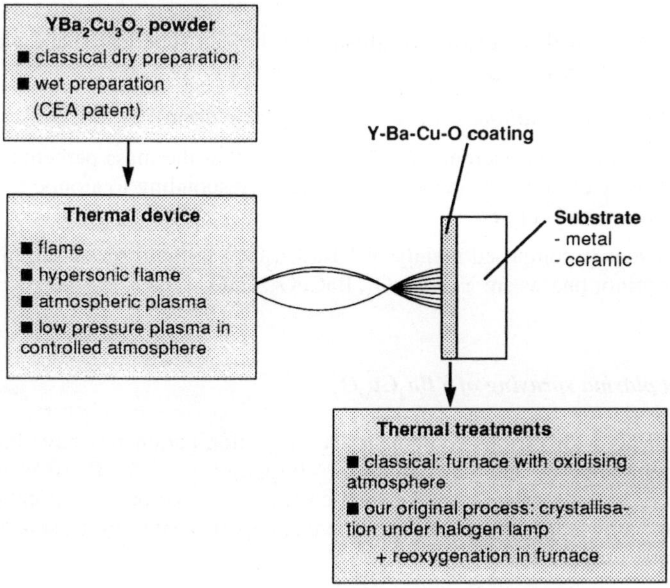

3-2 *The powder*

In most studies of thermal spraying, the $YBa_2Cu_3O_7$ powder is prepared by solid state diffusion (11, 13). The synthesis procedure is to make a high temperature reaction of a mixture of precursors, Y_2O_3, CuO and $BaCO_3$ or BaO_2 in the required stoichiometry. A series of operations of sintering and grinding is used to obtain particles made mostly of the phase 1, 2, 3 and size between 20 and 150 μm.

We have chosen in our case, to synthesize a powder using the wet method according to a procedure developed at the CENG (14). This enables a powder to be obtained with a better controlled granularity and a good homogeneity within each agglomerate.

The main stages of the preparation of this powder are as follows:

- dissolution of copper in nitric acid (formation of $Cu(NO_3)_2$),
- precipitation of copper using $Ba(OH)_2$ then with Y_2O_3
- formation of a slurry made of $Cu(OH)_2$, $Ba(NO_3)_2$, $Y(NO_3)_3$
- spray drying of the slurry

The powder is made of $Cu_2(OH)_3NO_3$, $Ba(NO_3)_2$ $Y_2(OH)_x(NO_3)_y$.

The following operations are then carried out:

- thermal treatment of the decomposition of nitrates and hydroxides at 750°C for 4 hours
- calcination at 920°C for 5 hours to make the phase $YBa_2Cu_3O_{7-\delta}$ from the precursors

Morphology of the powder

The particles with a porosity of about 50% have a spherical shape which gives the powder satisfactory flow properties.

Composition and crystalline structure (ion coupled plasma)

Emission spectra measurements (ICP) indicate that the mass percentages of the coatings were very close to the theoretical values corresponding to atomic ratios 1, 2, 3 for yttrium, barium and copper.

The powder is composed mainly of $YBa_2Cu_3O_{7-\delta}$ in the form of sub-micron crystals. The other minor phases are Y_2BaCuO_5, $BaCuO_2$, CuO.

3-3 *Hot plasma spraying of $YBa_2Cu_3O_7$*

Many trials with various equipment and projection parameters have been carried out for the preparation of ceramic coatings of $YBa_2Cu_3O_7$ (15, 17). Plasmas spraying seems to be the most appropriate to make such coatings, and the best electrical characteristics are, to our knowledge, those measured on coatings deposited by plasma spraying under reduced pressure (17).

The plasma carrier gases most often used are argon (18), nitrogen (9) and helium or Ar-He or Ar-H_2 mixtures (20-21). The role of the hydrogen is to increase the heat transfer from the plasma to the particles, that is to encourage fusion. However, the addition of a small quantity of hydrogen (1 l/mn) can cause overheating of the particles (beyond the peritectic temperature) and chemical segregation which can persist after the quenching of the particles on the substrate (10^6 K/s). Hydrogen can also irreversibly remove oxygen from the lattice of the perovskite structure.

Hydrogen therefore reduces the quantity of superconducting phase as has been observed by Neiser in his samples (20), but it is also important to note that even without the addition of hydrogen, loss of oxygen occurs during the plasma spraying.

The conditions for plasma spraying that we have adopted after optimisation of the procedure are given in another publication (22) and are reviewed in table II.

The substrates used in the framework of theses trials are made of stainless steel, chosen for its high thermal stability and thermal expansion coefficient close to that $YBa_2Cu_3O_7$ In addition, to reduce the problems of interdiffusion during the thermal treatment, "post spraying", and to improved the adhesion of the coating, an undercoat of NiCrAlY was systematically sprayed on the substrates.

Plasma gas discharge(l/h)	Carrier gas discharge(l/h)	Power (KW)	Spraying distance (cm)	Substrate temperature °C
3250 (Argon)	200 (Argon)	22	7	40 (cooling by convection)

Table 2: Atmospheric projection parameters

After spraying, the deposits of YBaCuO are poorly crystallised, oxygen deficient and must be given a thermal treatment to restore the superconducting properties

3-4 The post-spraying thermal treatment

The thermal treatment has 2 roles :
– to recrystallise the superconducting phase
– to oxidise the samples after the oxygen loss during the sprayings

Thermal treatment in a furnace

The heat treatment is generally carried out in a furnace in air or oxygen. There are two main steps:

– a high temperature plateau (950-970°C) in oxygen (23) or (900-920°C) in air (24) to recrystallise the $YBa_2Cu_3O_{7-\delta}$ phase. The length of this plateau is between 1/2 h and several hours

– a cycle at a lower temperature (around 500°C)

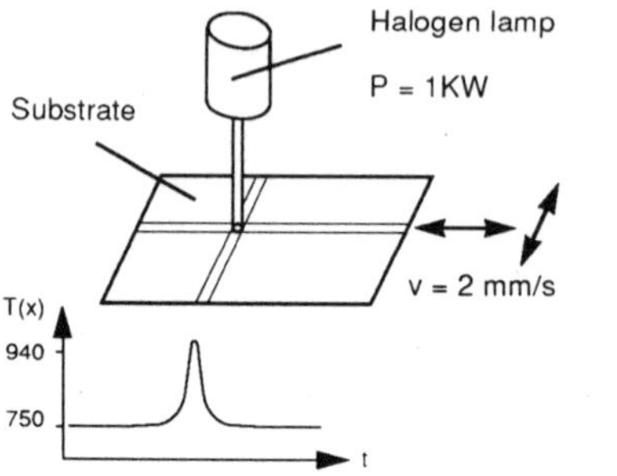

Figure 1: Principle of heat treatment under halogen lamp

Thermal treatment under a halogen lamp

To limit interdiffusion and maintain the substrate at lower temperature during the heat treatment. We have developed a halogen lamp heating system (1 KW power).

The coating to be heat treated is displaced under the halogen lamp beam (a spot of a few mm^2) by an automatic system. Each part of the coating is then subject to a temperature of 940°C (when under the beam) and to a temperature of 750°C when out of the beam (see fig. 1).

After this phase which recrystallises the coating, the sample is heated in a furnace to 500°C in oxygen (reoxidisation of the $YBa_2Cu_3O_{7-\delta}$ phase).

Comparison of samples treated in a furnace or under a halogen lamp

Two coatings prepared identically by atmospheric plasma spraying were heat treated in similar conditions of temperature (930°C) and duration (1 hour plateau) in a furnace and under a halogen lamp. Electron microprobe analysis was used to reveal the diffusion of the elements in the two samples. The reaction between the coating and the yttrium rich underlayer leads to the formation of a green phase as shown by X-ray diffraction of a coating detached from its substrate. This green phase forms an insulating layer between the deposit and its substrate. In the case of the halogen lamp heat treatment, this insulating layer is very thin (< 1 μm). For the sample heated in a furnace, the reactive zone is much thicker and can reach several tens of microns (table 3).

Sample before thermal treatment	**Sample after thermal treatment in a furnace**	**Sample aftert thermal treatment under halogen lamp**
YBaCuO 250 μm Undercoat NiCrAlY 200 μm	$YBa_2Cu_3O_{7-\delta}$ Y_2BaCuO_5 (up to 100 μm) phase interdiffusion zone	YBa$_2$Cu3O$_{7-\delta}$ Y_2BaCuO_5 (a few μm)
Substrate 2 mm	Substrate 2 mm	Substrate 2 mm

Table 3

4 - CHARACTERISTICS OF THE $YBa_2Cu_3O_{7-\delta}$ COATINGS

4-1 Chemical composition

During the plasma spraying, the chemical composition of the particules of powder heated to a high temperature (beyond the fusion temperature) can change. The phenomenon can occur:

– when the plasma power is to high, which leads to preferential sublimation of certain constituents

– when reactions between the particles and the gas take place, which lead to melting of the powder.

a) Loss of cations Y, Ba, Cu

We have measured the cation content by emission spectroscopy (ICP) of the basic powder and on coatings to determine if such composition changes occur during spraying (see table 4).

Mass %	Measurement counting error	Theoretical %	Powder	Atmospheric plasma
Y	1 to 2%	13.3	13.9	14
Ba	5%	41	35.7	39
Cu	1 to 2%	28.6	28	29

Table 4: Cation content (Y, Ba, Cu) of the basic powder and a plasma coatings

We can see that the spraying conditions are very well adapted since the mass percentages of the cations (Y, Ba, Cu) of the coatings are very similar to those of the basic powder.

b) Oxygen loss

Oxygen losses occur during the plasma spraying $YBa_2Cu_3O_7$ These are related to the fusion of the material and to the low partial pressure of oxygen around the particles sprayed. The oxygen loss during plasma atmospheric spraying measured by thermogravimetric analysis is 2.5% by weight.

After the thermal treatments for recrystallisation under the halogen lamp and for reoxidation in a furnace, the $YBa_2Cu_3O_{7-\delta}$coatings have an oxygen content equivalent to $\delta = 0.1$ (iodometric measurement).

4-2 Crystalline structure

After plasma spraying of $YBa_2Cu_3O_7$, the coatings are made up of the mixture of an amorphous phase and microcrystalline particles. The phases identifiable by X-ray diffraction are: Y_2O_3, Y_2BaCuO_5, $YBa_2Cu_3O_{7-\delta}$.

After thermal treatment the phases determinated by X-ray diffraction (see fig. 2) are :
- $YBa_2Cu_3O_{7-\delta}$ major phase
- Y_2BaCuO_5 and CuO very minor phases with intensities less than the basic powder

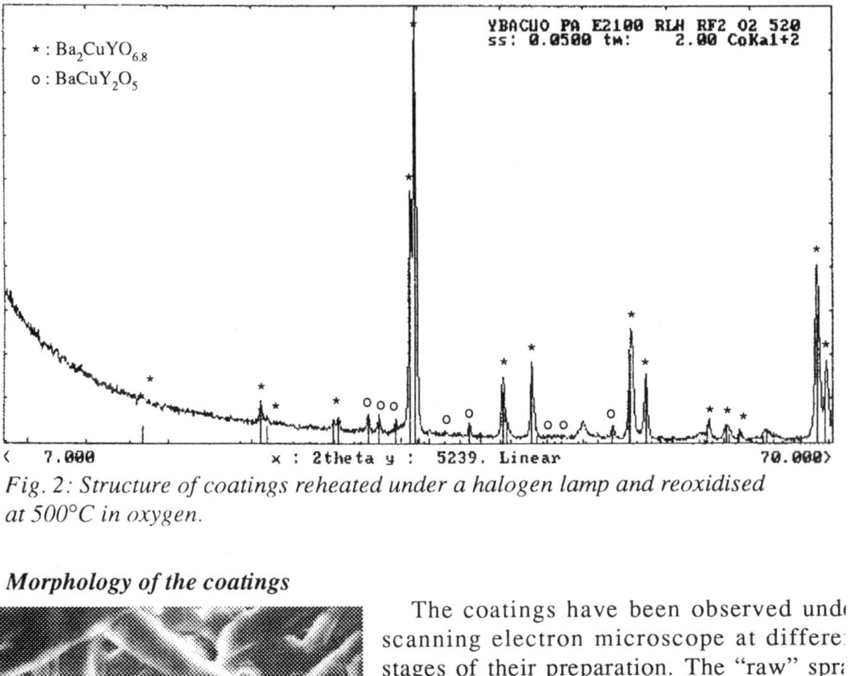

Fig. 2: Structure of coatings reheated under a halogen lamp and reoxidised at 500°C in oxygen.

4-3 Morphology of the coatings

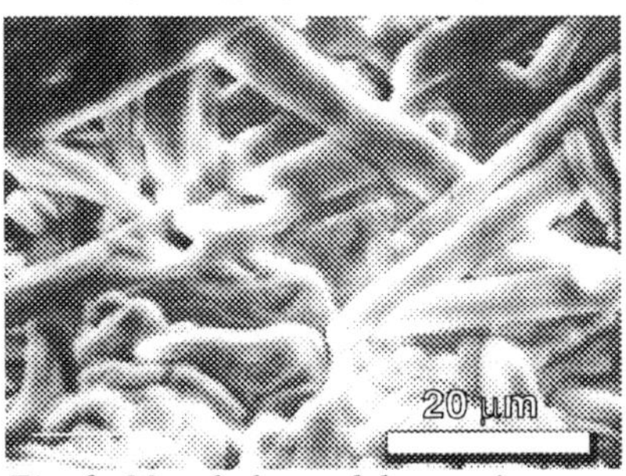

Fig. 3: Morphology of the coatings.

The coatings have been observed und(scanning electron microscope at differe stages of their preparation. The "raw" spr: coatings appear to be made up of adjacent me ted particles and have a lamellar character.

After reheating, the recrystallisation visible by the formation of rods of the sup(conducting phase (see fig. 3).

The platelets of the superconducting ph: have sizes between 2 and 15 µm (mean len; 11 µm, mean thickness 3 µm).

The green phase (Y_2BaCuO_5) was identified by X-ray microanalysis as grains : small clear particles about 2 µm in size. In addition, figure 3 shows the porosity of coatings and the presence of cracks. Porosity measurements were made by alcohol absorption and by image analysis (see table 5).

These values are quite normal for plasma sprayed ceramic coatings.

	total porosity %	open porosity	closed porosity	roughness (Ra μm)
plasma sprayed $YBa_2Cu_3O_7$	12.6	4.5	8.1	12

Table 5 : Porosity and roughness of coatings made by atmospheric plasma spraying

4-4 Electrical and magnetic characteristics

The interest in superconducting materials is due to their electrical characteristic (zero electrical resistance) and in their magnetic properties (perfect diamagnetism). Applications of these materials require high critical current densities at temperatures and magnetic fields as high possible.

Pure coatings of $YBa_2Cu_3O_7$ prepared by thermal spraying are generally polycrystalline and untextured. The critical current densities are unfortunately quite low, and are not suitable for power applications.

The limitation of the transport electrical current occurs in the grain boundaries since these can correspond to good or bad electrical contact zones.

Table 6 reviews the electrical results found in the literature. One should nevertheless note that most of these publications mention only the transition temperature.

For our study, systematic measurements of electrical characteristics and especially critical current densities have been used to optimize the preparation conditions; choice of the powders, determination of the spray parameters, development of thermal treatments.

Critical current density measurements were made by a device using pulsed currents [32]. This technique limits the heating near the connections and enables direct measurement of the critical current passing through the sample.

Electrical characteristics

The main electrical characteristics of $YBa_2Cu_3O_7$ coatings are shown in table 7:
– The critical temperature, $Tc(R = 0) = 90.5$ K is high and the transition length is very low

– The value of the resistivity (0.7-0.8 mΩ.cm) at 300 K, very low for this type of sample, is close to that of the best sintered samples

Finally, the values of the critical current density particularly high (1000 A/cm^2) are the best values published in the literature.

References	Tc Beginning (K)	Tc (R=0) (K)	Magnetic screening	Critical current density A/cm²	Type of spraying
R.A. Neiser J.P. Kinkland H. Herman (25)	88 (mean)	81	10% to 35 K 200 G	< 100	atmospheric plasma
K. Tachikawa I. Watanabe S. Kosuge (26)	96			690 (77 K, 0T)	low pressure plasma
J. Karthikeyan K.P. Sreekumar (27)	91			3 to 15	atmospheric plasma
C.M. Liao X.I. Chen X.W. Qi D.M. Du (28)	86,8	78,5			flame
A.M. Bouanani R. Suryanararyanan (29)	92,5			70 (77 K, 0T) - 10^5 (10 K, H < 1 T)	atmospheric plasma
T. Konaka L. Sankawa T. Matsuara (30)	94	88		6 (77 K, 0T)	atmospheric plasma
W.F. Chu F.J. Rohr (31)		87		270 (77 K, 0T)	atmospheric plasma

Table 6

Critical temperature Tc (R = 0)	Transition width ΔTc (90%-10%)	Resistivity ρ(300 K) (mΩ.cm)	ρ(300 K)/ ρ(100 K)	Critical Current Density (A/cm^2)
90.5 K	1 K	0.7-0.8	2.35	1000

Table 7: Electrical characteristics of $YBa_2Cu_3O_7$ coatings made by atmospheric plasma spraying (CEREM)

Magnetic characteristics

The screening properties of $YBa_2Cu_3O_7$ coatings have been evaluated and compared to those of classical materials such as copper or Fe-Si 3% high-permeability alloys.

In addition, in order to avoid interpretational errors and to evaluate the behaviour of the coatings alone, tests were also made on substrates with the NiCrAlY coatings only. Alternating magnetic field screening measurements [33] show that at 77 K for example, for frequencies less than 1 KHz, the screening efficiency of $YBa_2Cu_3O_7$ coatings is better than that of copper or Fe-Si 3% alloys currently used as magnetic screening.

Conclusion

The results published in the literature confirm the possibility of making thick coatings (50 μm) to several mm) of $YBa_2Cu_3O_7$ by plasma spraying on substrates of various materials and shapes.

The final quality of the coatings depends on all the links of the chain such as:

- preparation of the powder
- spraying
- thermal treatment of the coatings.

Optimisation of the preparation process requires extensive characterisation corresponding to each stage of the preparation (composition, structure, microstructure, electrical properties).

The best results (critical current density Jc = 1000 A/cm^2 at 77 K, 0T) are obtained by atmospheric plasma sprayed coatings, heat treated for recrystallisation (halogen lamp) and for reoxygenation (500°C furnace) in two stages.

The use of a halogen lamp system enables preferential heating of the coating while limiting the temperature of the substrate, so as to reduce the diffusion phenomena which can lead to destruction of the superconducting phase.

Good electrical properties of these coatings as well as ease of use of the hot spraying techniques (automated systems, coating speed of 1 to 2 kg/hours) allow us to envisage, in the short term, the production of large sized parts which can be brought to industrial scale in such fields as magnetic screening for biomedical (low frequencies) or military application. Tests are currently taking place.

Finally, power applications could be envisaged in the future if the critical current density can be increased. One promising method would seen to be melt texturing of coatings.

REFERENCES

(1) R.J. Cava, B. Batlogg, R.B. Van Dover, D.W. Murphy, S. Sunshine Phys. Rev. Lett 58, 1976 (1987)

(2) Y. Lepage, W.R. MC Kinnan, J.M. Tarascon, L.H. Greene, G.W. Hull et al.., Phys. Rev. B35, 7245 (1987)

(3) T. Siegrist, S. Sunshine, D.W. Murphy, R.J. Cava, S.M. Zahinak Phys. Rev., B35, 7137 (1987)

(4) A. Santoro, S. Miraglia, F. Beech, S.A. Sunshine, D.W., Murphy et al.., Mat. Res. Bull., 22, 1007 (1987)

(5) M.O. Eatough, D.S. Ginley, B. Morosin, E.L. Venturini Appl. Phys. Lett 51, 367 (1987)

(6) M. Kogachi, S. Nakanishi, K. Nakahigashi, S. Minamigawa Jpn- J. Appl. Phys. 27 L1228 (1988)

(7) Y. Enomoto, T. Murakami, M. Suzuki, K. Moiwaki Jpn J. Appl. Phys. 26, L1248 (1987)

(8) P. Regnier, X. Deschanels, L. Chaffron 2de Journées d'étude sur les céramiques supraconductrices à haute temperature critique 6 et 7 novembre, 181 (1990)

(9) RS. Roth, K.L. Davis, J.R. Dennis Adv. Ceram. Mat. 2, 303 (1987)

(10) M. Gervais, A. Dany, B. Dubois Rev. Phys. Appl. 24, 495 (1989)

(11) W.T. Elom, J.P. Kirkland, R. Neiser, E.F. Skelton Adr. Ceram. Mater, 2, 411 (1987)

(12) C.R. Guarnière, R.A. Roz, S.A. Shinvashankar, J.J. Cuomo J of Vacuum Si. and Techn. (nov. 1987)

(13) S. Rohr, G. Eckart, I. Bacher, V. Schlafer Modern Phys. Lett. B4, 717 (1990)

(14) Brevet CEA, Dehandt (1990)

(15) J.J. Cuomo, C.R. Guarnieri, S.A. Shimvashankon, R.A Raz Adv. Ceram. Mater. 2, 422 (1987)

(16) P. Mc Go,, N. Jain, Thin solid films, 166, 163 (1988)

(17) K. Tachikawa, M. Ono, Y. Shimbo, T. Suzuki, M. Kabasawa I.E.E.E. Trans on magn., 25, 2029 (1989)

(18) C. Moreau, L. Parent, S. Dallaire B. Champagne Surf. Coat. Technol., 37, 305 (1989)

(19) L.S. Wen, S.W. Qian, Q.Y. Hu, H.W. Zhao, K. Yang
Thin solid films, 152, L143 (1987)

(20) R.A. Neiser, J.P. Kirkland, W.T. Elam, H. Herman
Mat. Res. Symp. Proc. 99, 689 (1988)

(21) Y. Arata, A. Ohmori, S. Sano
Proc. Therm. Spray. Conf. A.S.M. Cincinatti, 183 (1988)

(22) J. Lacombe, J. Danroc, G. Kunka
J. of the Less. Com. Mat., 164, 509 (1990)

(23) P. Me Ginn, N. Jain, D. Lee
Surf. and coatings tech. 37, 359 (1989)

(24) S. Kitahara, Y. Yoshida, T. Fukushima, S. Kuroda
Conf. London Symp. P 86-1 (1988)

(25) R.A. Neiser, B. Gudmundoson, Y.M. Zhu, H. Herman
Int. Thermal spray conf. London, June 4-9 (1989)

(26) K. Tachikawa, I. Watanabe, S. Kosuge, M. Kabasawa
Appl. Phys. Lett. 52, 1011 (1988)

(27) J. Kaathikeyan, K.P. Sreekunan, M.B. Kurup
J. Phys. D : Appl. Phys. 21, 1246 (1988)

(28) C.M. Liao, X.L. Chen, X.W. Qi, D.M. Du
Conf. London Symp. P20-1 (1988)

(29) A.M. Bouanani, R. Suryanarayanan, G. Bum, O. Gorochov
2de Journée d'étude sur les céramiques supraconductrices à haute température critique 6 et 7 novembre, 215 (1990)

(30) T. Konaka, I. Sankawa, T. Matsura, T. Higashi, K. Ishihara
Jpn, J. Appl. Phys. 27, L1092 (1988)

(31) W.F. Chu, F.J. Rohr
Physica C 153, 802 (1988)

(32) F. Weiss
Laboratoire des matériaux et du génie physique (INPG)

(33) Thèse J. Lacombe
Thèse de docteur ingénieur de l'INPG (1991)

Chapter 12

PLASMA SPRAY COATINGS OF HIGH Tc YBaCuO SUPERCONDUCTORS

G. Swaminathan, K. Venugopal, D. Maheswara Reddy,
K. A. Durgaprasad and R. Somasundaram
Corporate Research & Development Division
Bharat Heavy Electricals Ltd.
Vikasnagar, Hyderabad-500 593, INDIA

Abstract

There has been a concentrated effort to improve the superconducting properties in high Tc superconductors which will enable their widespread utility. The processing of the high Tc superconductors to make them in usable form is equally challenging. Plasma spraying of high Tc superconductors offers some scope for large scale processing. The results and observations on plasma sprayed YBaCuO coatings are presented in this article.

1. Introduction

With the announcement of the discovery of a 93°K transition temperature in Y-Ba-Cu-O system by Wu et al [1], the race to develop viable technologies to process these brittle ceramic superconductors into commercially exploitable forms has become intense. Extensive work has been carried out in processing the high temperature superconductors into commercially useful forms, such as, wires and tapes [2-12]. But rapid technological progress has been hampered by the fact that, so far we have not been able to find a satisfactory guiding principle for obtaining materials with high critical temperature and critical current densities. For instance, can we expect high Tc superconductivity only in materials having Cu-O planes in their crystal structure? If the answer is yes, how many Cu-O planes? A recently reported thallium compound containing four Cu-O planes in a unit cell did not show higher Tc as expected, belying the hitherto held belief that more Cu-O planes per unit cell would result in higher Tc. Again, does Tc rise with the degree of complexity in the crystal structure? Why does a small amount of oxygen deficiency affect the critical temperature significantly? The theory of high Tc superconductivity will hopefully provide answers to many of these technologically important questions in the very near future. Meanwhile, there are many hurdles which need to be crossed before high Tc superconducting materials can be put to meaningful use. They include, increasing the critical current density (Jc), sustaining high current densities even in the presence of high magnetic fields, making long lengths of wires and tapes which are very important for high power applications, improving the mechanical properties and stability under prolonged exposure to atmosphere, arriving at the best material composition, and finally, optimising the processing parameters for large scale manufacture. While very high current densities have been achieved in thin films and single crystals, large scale applications such as generation, transmission and storage of electricity will probably require the material to be produced in the form of wires or tapes.

In this paper, we will review the work done on plasma- sprayed coatings of Y-Ba-Cu-O superconductors. In spite of the fact that only moderate current densities (~ 400 to 600 A/cm^2) have so far been achieved using this technique, many groups are pursuing this route for the following reasons. Plasma spray technique offers advantages of high mechanical strength and durability, high rates of deposition over large areas, applicability to a wide range of coating and substrate materials, good adhesion, flexibility in producing a wide variety of geometries and thicknesses and its potential for automation and continuous manufacture. The Y-Ba-Cu-O system was chosen for extensive investigation because they offer some obvious advantages. The starting raw materials for this system are non-toxic and hence they score over thallium based compounds when it comes to routine large scale handling of powders. They are much less volatile compared to Bismuth based system thereby offering much better control over processing.

2. Plasma Spray

2.1 Powder Preparation and Characterisation

The Y-Ba-Cu-O powder which is to be used as a feed stock for plasma-spraying, is prepared by grinding stoichiometric mixture of high purity Y_2O_3, $BaCO_3$ and CuO in a planetary ball mill. The resultant mixture is subjected to solid- state reaction at 900°C in air for 24 hrs followed by cooling in air at the rate of 50°C hr^{-1}. The calcination process is usually repeated thrice to obtain homogeneous powder. The powder is usually prepared in a batch of 500 gms. The resultant material is ground and sieved to obtain powders with particle size in the range 44-90 μm. In case of powders with additives like Ag_2O, SnO_2 and Bi_2O_3, require quantities of these additives were mixed with YBaCuO powders and ground in a ball mill prior to spraying.

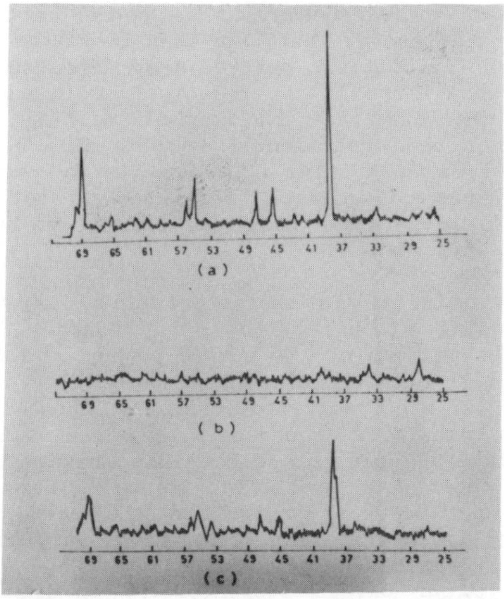

FIG.1 X-RAY DIFFRACTION PATTERNS OF a: SOURCE POWDER b: AS SPRAYED FILM c: FILM AFTER HEAT TREATMENT

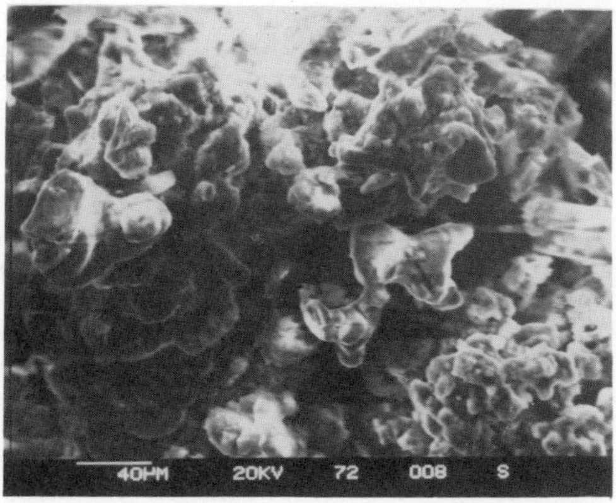

FIG.2 SEM MICROGRAPH OF THE POWDER USED FOR SPRAYING

In the plasma-spray processing, the powder particles are imparted tremendous amount of energy that take them to molten state even as they travel at high velocities before they are subjected to rapid solidification when they hit the relatively cold substrate. Given our experience, we believe that incongruent melting of non-spheroidal particles has not been detrimental to obtaining good quality, dense depositions of Y-Ba-Cu-O films. Our results, vis-a-vis those of our contemporaries reported in the literature, do not suggest that the simplified powder preparation route followed by us has resulted in any technical trade off.

The source material is characterised by X-ray diffraction for phase purity, using a cobalt anode X-ray source. Fig.1(a) shows the X-ray diffraction pattern of the source powder, which is predominantly a single phase with orthorhombic defect perovskite structure. When the source powder is sintered into pellets, they invariably exhibit superconducting transition around 90°K. Cuomo et al [6] who used much finer powders (average particle size ~10μm) have estimated the actual composition of their source material to be $Y_{1.1}Ba_{2.1}Cu_3O_{9.5}$ by optical emission technique. oxygen content in the powder used by us was estimated, using Iodimetric titration, to be varying between 6.75 and 6.80. However, it has been amply demonstrated, and understandably too, that phase purity or oxygen stoichiometry in the starting material (powder) is not an essential requisite for getting good Tc on sprayed films - as the material undergoes considerable phase change during plasma spray processing. Fig.2 is a typical SEM micrograph of the powders used routinely in our work for plasma spraying.

2.2 Plasma Spraying

The basic concept of plasma-gun operation is described in detail elsewhere in this volume . Essentially, in the plasma-spray process, feed stock material in the form of powder is injected into a high temperature high velocity plasma stream. The temperatures and velocities attained by these powder particles have been variously estimated to be of the order of 4000-12000°K and 100-200 m/s [6]. These fast moving molten droplets form a deposit (coating), when they are suddenly stopped by a relatively cold substrate, by deformation and rapid solidification. We can therefore see that heat and momentum transfer from the plasma jet to the powder particles are of crucial concern for achieving good coatings. Therefore arc current, plasma gas composition and flow and particle size in the powder are the primary parameters which require to be optimised. Substrate distance from the gun nozzle and substrate temperature are also important parameters as they dictate the speed with which solidification takes place on the substrate and hence porosity and microcracks in the deposited film.

Basically three types of plasma spray processes were attempted by

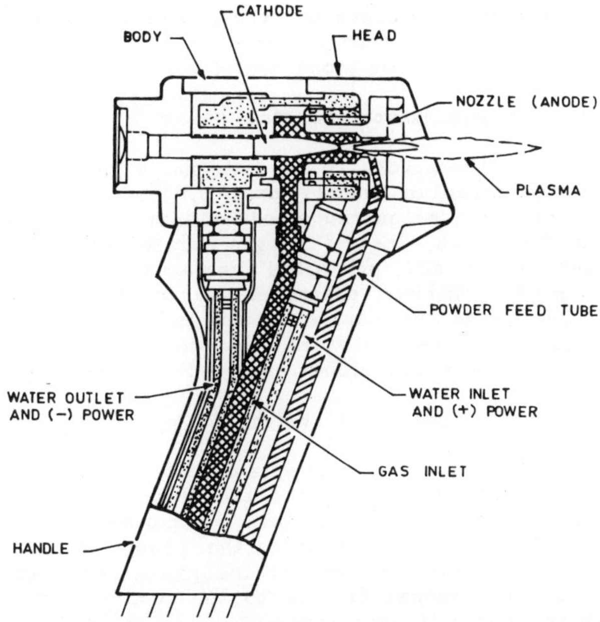

FIG.3 CROSS SECTION OF INTERNAL SPRAY GUN

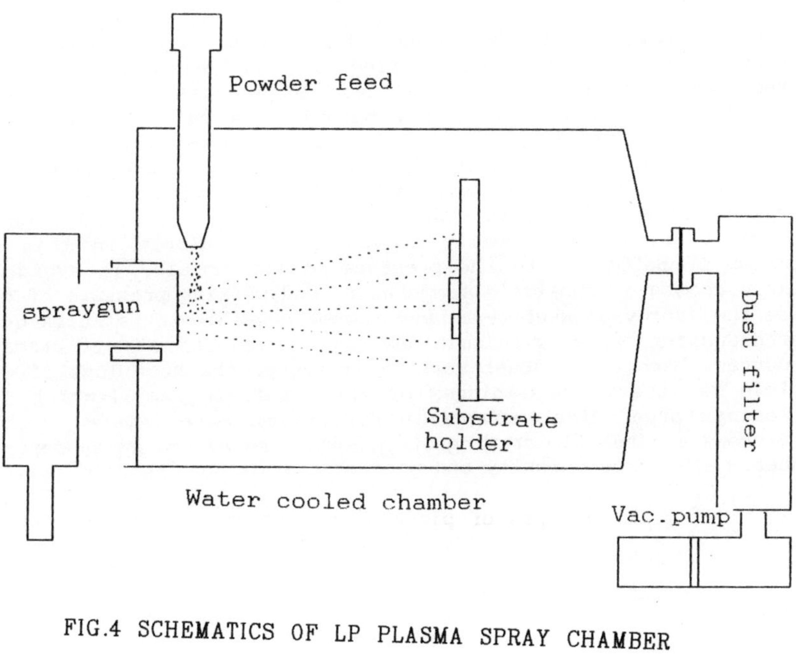

FIG.4 SCHEMATICS OF LP PLASMA SPRAY CHAMBER

us, namely, internal feed plasma spray, external feed plasma spray, and low pressure plasma spray systems. Each of these systems are briefly described in the following sections.

2.2.1 Internal feed plasma spray system

In this system the powder is fed internally into the plasma flame. Here a Bay State Abrasive Plasma Spray System of 40 KW capacity is used. In this system, the powder feeding unit, introduces the powder directly into the plasma, inside the gun-nozzle, where it is heated to a near molten state and projected as a spray. Fig.3 shows the cross-section of the internal plasma spray gun and the location of the powder feed tube inside the gun nozzle. During the internal feed plasma spraying process, the resident time for the powder particles is much longer compared to the conventional external feed plasma spray system.

2.2.2 External feed plasma spray system

This is conventional plasma spray coating system in which the powder is externally fed from outside the plasma gun. 9MB Metco Plasma Spray System is used for this purpose. A powder feed unit supplies regulated quantities of powder into the plasma flame, outside the gun-nozzle. Argon is used as the working gas. Argon gas shielding is also provided by an external fitting to the gun, in order to create an inert atmosphere around the substrate.

2.3 Low Pressure Plasma

So far we have described the plasma spray coatings of superconducting YBaCuO films carried out at atmospheric pressure. Another technique tried by us for these depositions was "Low pressure plasma spraying". Here the plasma spraying is carried out inside a chamber which is evacuated and maintained at a lower working pressure, namely 50-200 torr, by filling the chamber with a desired gas. This system affords some obvious advantages like deposition process strictly under controlled atmosphere, and better quality depositions both in terms of density and impurity contents. Particularly in this case, formation of $BaCO_3$ due to the presence of CO_2 in air, is avoided. By suitable choice of substrate temperature and working pressure of oxygen inside the 'spray-chamber' - it may become possible to get high-quality superconducting film with minimum post deposition heat treatment procedures. Because of lower working pressure, the heat loss of molten particle is less as it impinges on the substrate facilitating higher flattening force. This results in denser and more layered coatings. Fig.4 shows a schematic drawing of low-pressure plasma spray deposition chamber.

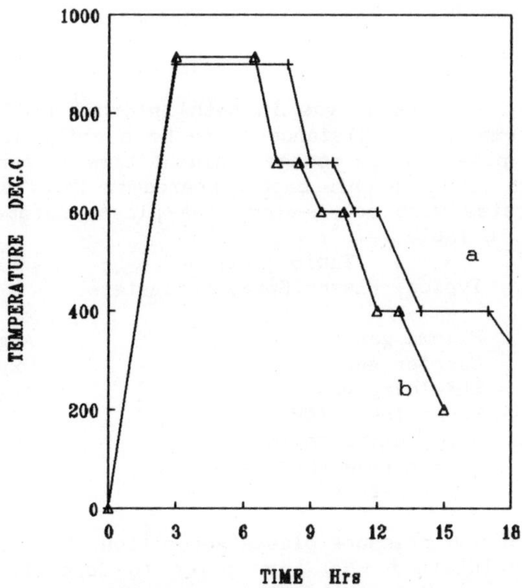

FIG.5 OPTIMISED HEAT TREATMENT CYCLE FOR
a: INTERNAL b: EXTERNAL SPRAYED COATINGS

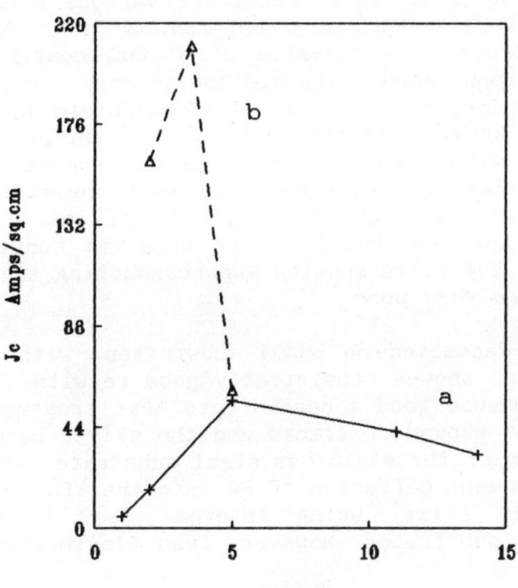

FIG.6 EFFECT OF SINTERING TIME ON Jc FOR
a: INTERNAL b:EXTERNAL SPRAYED COATINGS

3. Processing of YBaCuO

3.1 Spray Parameters

Argon is used as working gas in both internal and external feed plasma spray systems. Spray distance could be a critical parameter in the external feed plasma spray as the resident time of particles in the plasma jet is much lower in this case compared to internal feed plasma spray. The substrates were not heated. The plasma parameters used in general are shown in Table 1.

Table 1:
Typical Plasma Spray Parameters

Plasma gas	Argon
Carrier gas	Oxygen
Shielding gas	Argon
Power level (KW)	20
Spray distance (mm)	80
Powder feed rate (g min^{-1})	17
No of Passes	6

In the case of low pressure plasma deposition, the vacuum level in the chamber was maintained <200 torr prior to deposition. 9MB Metco plasma spray gun was used in this process.

3.2 Substrates

While narrowing on an ideal substrate various substrates such as SS304 and Inconel were tried. Bond coats such as Ni-Cr, Ni_3Al, Cu and Ag were tried to increase the adhesion of YBaCuO coatings to the base substrate. Conventional bond coats did not improve the adhesion and the coatings became powdery or adhesion was poor. Cu did not lend itself to heat treatment at 900°C as it tended to oxidise during heat treatment. When YBaCuO films were directly deposited on inconel substrates, the resultant films, after heat treatment, showed Fe peaks in their XRD and many cracks could be observed in the films. As a result, superconducting transition could not be observed consistently in all the films. Even in the films showing superconducting transition, the Jc values measured were very poor.

YBaCuO films deposited on SS304 substrates, with a bond coat of wire-sprayed silver, showed consistently good results. The silver bond coat was able to ensure good adhesion upto heat treatment temperatures of 900°C. The films showed no cracks and the silver bond coat was able to prevent reaction of the stainless steel substrate with the deposited film and the consequent diffusion of Fe into the YBaCuO film. In fact, most of the early films, using internal feed plasma spray were deposited on these substrates. However, less flexibility and inability to obtain current density (Jc) values beyond 50-75 A/cm^2 compelled us to opt for 1mm thick cold-rolled silver strips as substrates. Apart from excellent flexibility, silver formed a good interface with the sprayed YBaCuO films. Silver also provides the cryostatic stabilisation which is very crucial for current density measurements at cryogenic temperatures. This idea was borrowed by us from our experience with measurements on low temperature superconductors. In addition, we were guided by the reports of very good improvements in mechanical properties.

3.2.2 Substrate preparation

SS304 substrates used were typically 10mm wide and 1.5mm thick and the length of the strips was no constraint. Typically, 50 to 100cm long strips were degreased, grit blasted to provide clean rough surface to promote good adhesion, and given a bond coat of silver by wire-spray technique. In the case of cold rolled silver substrates only degreasing was required as adhesion between silver and YBaCuO films is not dependent on mechanical lockup of these materials. These silver strips were typically 5mm wide, 1mm thick and 100cm long. For subsequent studies, however, these strips were cut into smaller length, after plasma spraying.

3.3 Heat Treatment

The samples, after spraying, are immediately transferred to a furnace maintained at 600°C and are allowed to cool slowly to room temperature. This procedure is necessary to prevent quench-cracking of the samples immediately after spraying.

The as sprayed coatings are poorly crystalline as can be seen from XRD pattern, shown in Fig.1(b) and are not superconducting. To recover the superconducting phase in the sprayed films, it is essential to subject them to appropriate heat treatment. The recovery of superconducting 1-2-3 phase after suitable heat treatment is shown in Fig.1(c).

The heat treatment conditions were optimised, initially, for internally fed plasma spray coatings and were later validated for other types of plasma spray coatings. The heat treatment temperatures were varied between 850- 950°C in flowing oxygen environment and the soaking time varied upto 11 hrs to arrive at an optimum time-temperature schedule. It became clear that cooling cycle is as important as the soaking time at the maximum temperature, for effective absorption and retention of oxygen by the YBaCuO films. The samples were cooled slowly and intermediate soaking at different temperatures was carried out to ensure full balance of oxygen stoichiometry during tetragonal to orthorhombic phase transformation. This is very essential as these

Fig. 7 MICROGRAPHS OF SPRAYED SAMPLES
a: INTERNAL SPRAYED COATING (SEM)
b: EXTERNAL SPRAYED COATING (SEM)
c: EXTERNAL SPRAYED COATING (OPTICAL)

tapes have a large surface area to volume ratio compared to bulk samples.

Samples heat-treated at 850°C exhibited multiple phases and were not superconducting. Heat-treatment at 950 °C resulted in partial melting of silver leading to peeling-off of the coatings.

Best results were obtained from samples heat-treated at 900°C for 5 hours, as per the heat treatment schedule shown in Fig.5. This heat treatment procedure was found to be optimum even for the coatings which contained additive oxides such as Ag_2O, SnO_2 and Bi_2O_3. The optimum heat treatment cycle for externally sprayed samples was 915°C for 3.5 hours in flowing oxygen. The effect of sintering time on the Jc values of these sprayed coatings at the optimised soaking temperatures are shown in Fig.6.

4. Results and Discussion

4.1 Composition and Physical Properties

External feed plasma spray coatings appear to be more uniform and of better quality upon visual observation. This is supported by the SEM micrographs of these samples after heat treatment as shown in Fig.7. The external spray samples 7(b) reveal very low porosity levels compared to the internal spray samples 7(a). The external spray samples show elongated and needle shaped grains suggesting preferential grain growth during heat-treatment. Also the twinning features could be observed in external spray samples as shown in the optical micrograph 7(c). Further, these samples show better adhesion to the substrate compared with the internal spray samples when subjected to bending to an arc of 30°. Internal spray samples showed a tendency to degrade upon exposure to atmosphere, indicated by the appearance of white and brown patches suggesting formation of barium carbonate and oxygen loss. In contrast, the external spray samples did not show signs of degradation even after exposure for a few months. Both the types of coatings were largely amorphous in as sprayed condition as revealed by XRD. X-ray fluorescence analysis showed that there is a copper loss during the internal feed plasma spray process. The percentages of Yttrium and Barium were, however, consistent with their composition in the original powder. Table 2 compares the percentages of copper in powders and sprayed samples at different power levels. It can be seen from the table that copper loss is quite pronounced at higher power levels and also in powders which contain 10% excess CuO. Copper loss was not observed in the external feed plasma sprayed samples. Therefore we can say that external spray samples are more consistent with respect to physical properties and chemical composition.

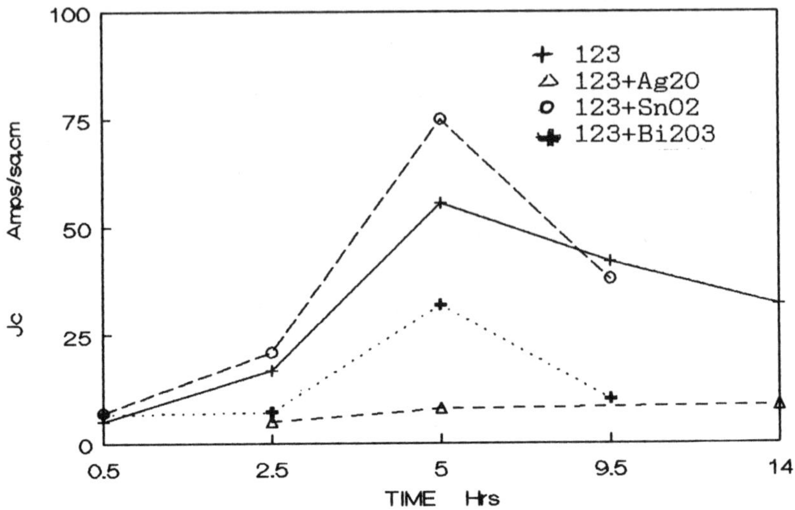

FIG.8 EFFECT OF ADDITIVES ON Jc FOR INTERNAL SPRAYED COATINGS

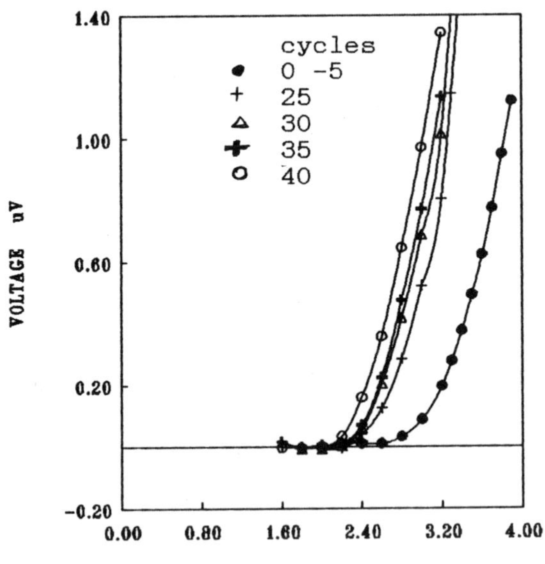

FIG.9 EFFECT OF RAPID THERMAL CYCLING ON RESISTIVE TRANSITION

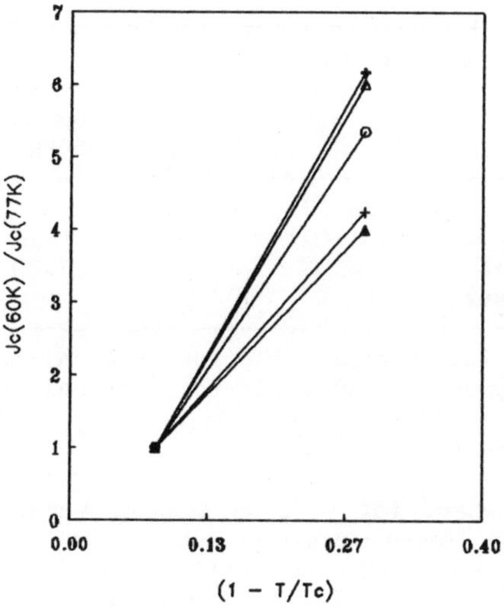

FIG.10 RATIOS OF CRITICAL CURRENT DENSITIES FOR NUMBER OF SAMPLES BETWEEN 60 K AND 77 K

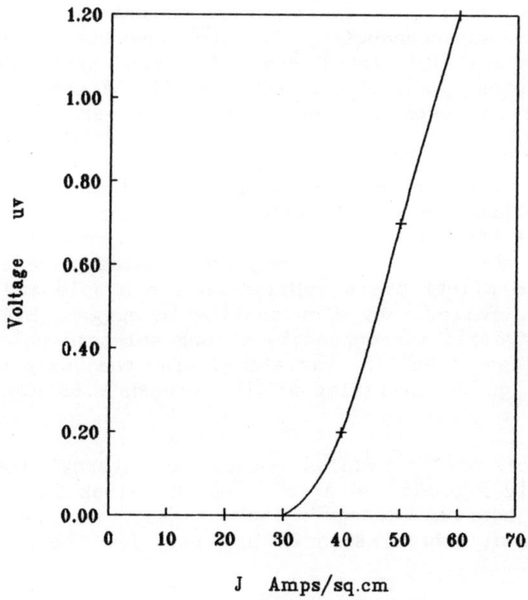

FIG.11 RESISTIVE TRANSITION CURVE FOR INTERNAL SPRAYED COATING

Table 2: X-ray Fluorescence for internal spray samples indicating copper loss

Material	Powder	Percentage of copper 20 kW	25 kW	30 kW
$YBa_2Cu_3O_x$	26.18	25.22	--	24.16
$YBa_2Cu_3O_x$ + 10% CuO	31.40	--	27.39	--
10% CuO added at preparation stage	31.47	--	27.00	--

Loss of yttrium and barium is negligible for internal spraying; loss of yttrium, barium and copper is negligible for external spraying.

4.2 Critical Temperature (Tc)

The Tc values measured for various samples were in the range of 82-84°K. The transition temperature was constant at 84°K in a typical experiment, irrespective of the soak time upto 15 hrs at 900°C. There was no measurable difference in Tc for the samples coated by internal feed and external feed plasma spray systems. Similarly the oxide additives, introduced to study their influence on superconducting properties of the YBaCuO films, did not influence the Tc.

4.3 Critical Current Density (Jc)

Critical current density is the maximum current density that can be sustained by the superconductor, in its superconducting state. Current densities beyond this level drive the conductor into normal resistive mode. The magnitude of Jc dictates the dimensions of the conductors and hence the economics of using the superconductors, for any given application.

Jc measurements were carried out using conventional dc four point probe measurement method and the Jc values were recorded at 1 μv/cm. For silver overcoated samples, contacts were made directly by soldering with 2% Ag-Pb/Sn solder. For other samples, contacts were made by soldering leads on to silver paste applied on the sample and fired at 450°C for 30 minutes. followed by slow cooling in oxygen. Samples were located in a glass cryostat surrounded by a long solenoid which ensures a uniform field (less than 1% variation) in the sample volume, facilitating studies on Jc variation in the presence of low magnetic fields.

The Jc values of YBaCuO samples coated by internal feed plasma spray technique rarely exceeded 60 A/cm^2. The Jc values for externally sprayed samples were however $200A/cm^2$. These results were obtained for optimised heat treatment schedules shown in Fig.6, for the internal and

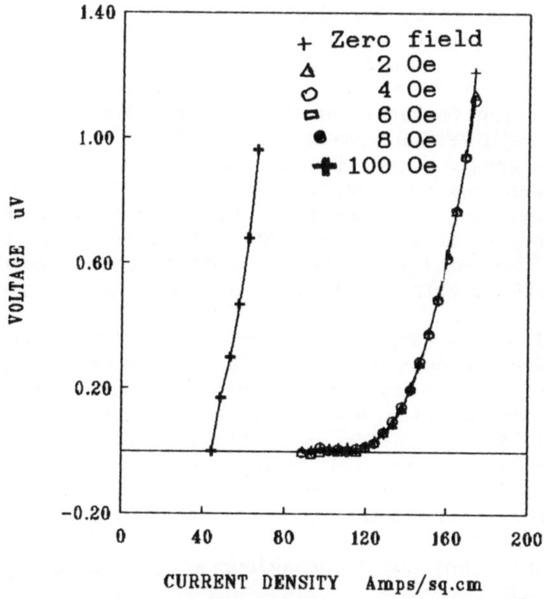

FIG.12 EFFECT OF MAGNETIC FIELD ON RESISTIVE TRANSITION FOR EXTERNAL SPRAYED COATINGS

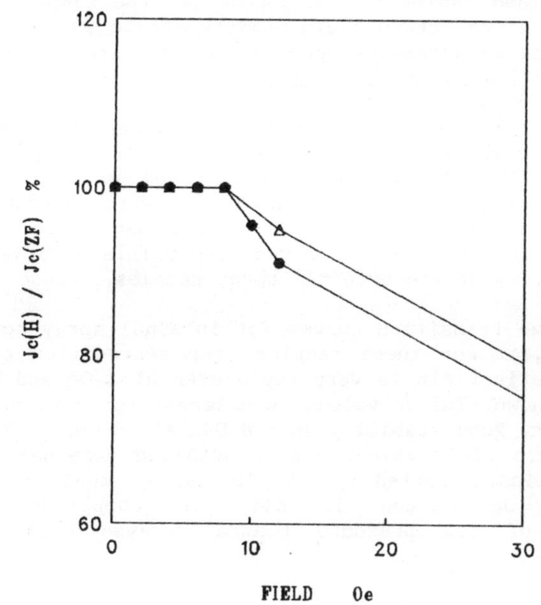

FIG.13 Jc-B CURVES OF EXTERNAL SPRAYED COATINGS

external feed plasma spray coatings. In an effort to improve the critical current density of the internally fed plasma spray coatings, the effect of additive oxides like Ag_2O, Bi_2O_3 and SnO_2 the YBaCuO powder used for spraying was studied. The results are summarised in Fig.8. Two interesting features can be observed. The Jc values for each of the additives with YBaCuO remain optimised for 5 hours soaking time at 900°C. This suggests that the internal spray process, more than anything else including the additives, is determining the quality of the final film and the 123 superconducting phase in them. Secondly, while the addition of 5% SnO_2 enhanced Jc from 60 A/cm^2 to 75 A/cm^2, other additives lowered the Jc values considerably, possibly due to widely varying melting points and hence increased porosity.

For internal spray samples, the Jc value falls drastically when cycled between room temperature and 77°K. The external spray samples show much better thermal cycling characteristics (Fig.9) with very small observable degradation even after a few cyclings between room temperature and 77°K. The ratios of critical current densities at 60°K to 77°K for a number of externally plasma sprayed samples are shown in Fig.10. The ratios are in the range of 4-6 as against 2 reported for bulk samples in the literature.

4.4 Resistive Transitions and Jc Behaviour at Low Magnetic Fields

For these measurements the samples were mounted on a fixture and oriented perpendicular to the magnetic field. Apart from the use of twisted and shielded cables for measurements, the samples were warmed up to room temperature before every measurement to avoid errors due to flux trapping. The measurements were carried out from zero field to 88 Oe. at 77°K as well as 60°K.

The critical current densities of internal and external feed plasma sprayed samples were typically 55 A/cm^2 and 160 A/cm^2 respectively at 77°K and zero field. Their n values, where n is defined as the slope of log V vs. log I plot and which characterises the effectiveness of inter grain coupling, were 8 and 11 for the internal and external plasma spray samples respectively. This once again confirms the superior quality of the external spray samples.

The resistive transition curves for internal spray coated samples is shown in Fig.11. For these samples, the degradation of Jc in the presence of magnetic field is very rapid even at 2 Oe and hence Jc (B) curves are not shown. The Jc values in external spray samples (Fig.12), however, show very good stability upto 8 Oe. At 88 Oe, Jc rapidly falls to 10% of its zero field value. A more detailed examination of the Jc (B) characteristics, plotted in Fig.13, shows that upto 8 Oe the degradation in Jc values is nil. In comparison the Jc(B) characteristics of low pressure plasma sprayed coatings are far

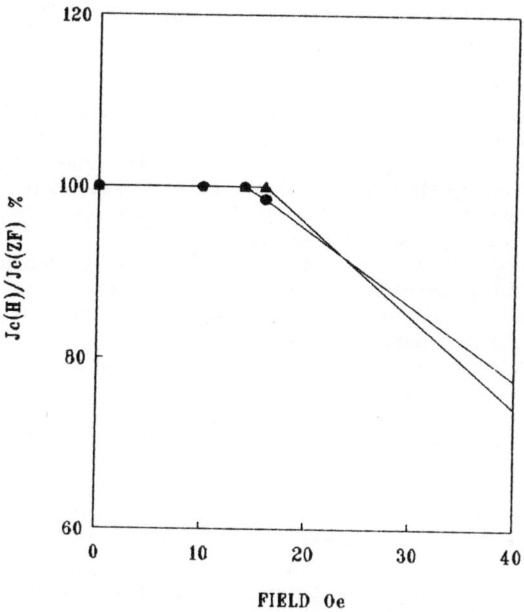

FIG.14 Jc–B CURVES FOR LP PLASMA SPRAYED COATINGS

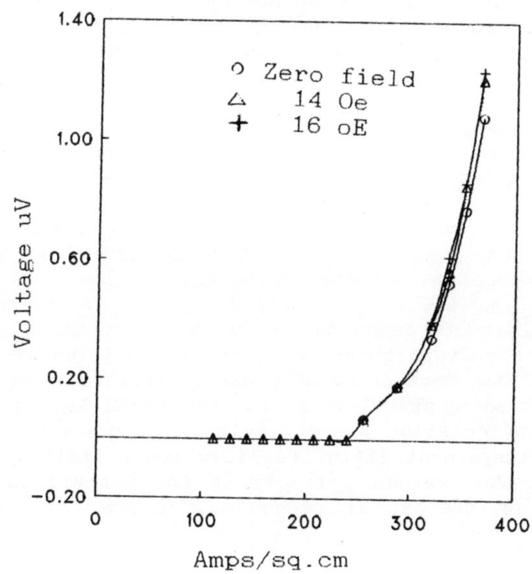

FIG.15 RESISTIVE TRANSITION CURVES FOR LP PLASMA SPRAYED COATINGS

superior(Fig.14). They exhibit identical transition upto 14 Oe(Fig.15).

4.5 Discussions

In the internal spray process, the resident time of the powder in the plasma is much higher in comparison to external spray process. This can be the cause of elemental loss in the internal spray process. Further, the aspect ratios of the particles also play an important role. There is preferential melting of the particles with large aspect ratios inside the gun nozzle. The molten particles resolidify when they come into contact with the nozzle wall. This results in the blocking of the nozzle and the consequent interruption in the spraying process. This obviates the possibility of spraying long lengths of tapes or building up film thicknesses greater than 200 μm. On the other hand, in the external spray process the resident time of the powder particles is determined by the spray distance only. Whether the particles are only in a plastic state or in a completely molten state will be determined by the plasma power and plasma composition. For example it is now established that for a given power, greater heating of the powders takes place if high enthalpy plasmas like Ar-H_2 are used compared to Ar-He plasmas. Since the powder is fed externally, the problem of nozzle blockage does not arise and uninterrupted flow of powder can be maintained. Thus, long lengths of tapes could be coated with-out interruption. It has also been possible to form dense and uniform coatings with thicknesses of 600 μm.

The superior quality of the external spray samples has been clearly brought out by all the measurements carried out by us. They have far better Jc values and n values indicating improved inter grain contact. Similarly they show much better thermal cycling characteristics .

The best Jc results, so far, for bulk polycrystalline YBaCuO samples have been of the order of $10^3 A/cm^2$. When it comes to plasma sprayed YBaCuO samples, we have not come across results of Jc exceeding 200 A/cm^2 for atmospheric plasma sprayed materials and 690 A/cm^2 for vacuum plasma sprayed materials. Reasons for these low Jc values can be traced partly to the microstructure of these materials and partly to the fundamental properties of the YBaCuO material.

The crystal structure and hence the superconducting properties of YBaCuO films are highly anisotropic. Current densities in the a-b plane are known to be 2-3 orders of magnitude higher at $10^6 A/cm^2$, compared to current densities along the C axis. In polycrystalline samples, random grain- to grain orientation would result in random orientation of the a-b planes. The current is therefore effectively limited . The problem is further aggravated by the porosity of the sprayed samples and the non-superconducting phases, if present. Although intra-grain current densities could be quite high, these interfaces, namely grain-boundaries, pores and non-superconducting phases, act as drains for the supercurrent. This is confirmed by the fact that poor quality films obtained by internal spray show Jc values of the order of 60 A/cm^2, while dense and uniform coatings obtained by external spray have already given us 200 A/cm^2 and samples(11) which are plasma sprayed under low pressure have already yielded 350 A/cm^2 (Fig.15). Clearly, therefore, it must be possible to obtain much higher Jc values with improved plasma processing condition and heat treatment procedures.

It is interesting to note that the ratio Jc(60°K)/Jc(77°K) is in the range of 6 for the best plasma sprayed (external) samples as against 2 reported for bulk samples. This is possibly due to better inter grain contacts and reduced junction dimensions at the grain boundaries.

The results obtained in externally sprayed coatings for fields up to 10 Oe indicate the absence of flux penetration and the complete meissner state of the samples. This could be due to negligible inhomogeneities and very small junction dimensions. The deposition technique seems to help the formation of layered structures, which prevent entry of flux across the samples. This is further confirmed by our measurements on Low pressure plasma sprayed sample showing complete meissner state up to 14 Oe. (Fig.14) The samples which show very good resistance to flux penetration up to 10-16 Oe also exhibit very good stability when subjected to rapid thermal cycling from room temperature to 77°K.

5. Prospects and Applications

5.1 Prospects

Plasma spray deposition technique has bright prospects for making thick superconducting films, suitable for many applications. From the foregoing results, it is clear that as we improve the quality of the polycrystalline films by refining the processing procedures, Jc values are bound to improve. Take for example in the case of vacuum plasma spray processing of the YBaCuO material, we have already doubled (nearly) the Jc values from 200 to 360 A/cm^2. Similarly, magnetic field shielding capability has been raised from 10 to 16 Oe. Values close to 700 A/cm^2 have already been reported [7]. The reasons are obvious; better control over spraying environment, much higher particle velocities and much denser coatings. A problem which cannot be easily overcome at the heat-treatment stage will be 6% volume contraction following Tetragonal to Orthorhombic transformation. Therefore the accent should be on reduction in the heat-treatment time and temperature, innovative approach to processing to obtain favorably oriented crystalline (123) as deposited coatings. Laser glazing is another prospective route as it could drastically reduce porosity, and alter the grain size in the films without inducing deleterious reactions at the substrate film interface. Low pressure plasma spray deposition of YBaCuO powder on to substrates heated to around 650 $\mathring{C}$ in an oxygen ambient, could possibly help us minimise the undesirable effects of post-spray heat treatment.

The future prospect of plasma spray technique lies in the fact that its potential for automation for carrying out spraying over long lengths, and over or inside massive objects has already been established in our laboratory. We expect that Jc values of the order of 1000-1500 A/cm^2 will be achieved in the near future using this technique.

5.2 Applications

The magnitude of Jc achievable presently and the Jc(B) characteristics limit the utility of these coatings. Fortunately for us, there are a few applications which do not need very high Jc values and our technology, as it is available today to make high Tc superconductors, could be put to use. Our results on LP plasma coatings where shielding upto 14 Oe has been obtained in a 150μm thick film as compared to 6 Oe/mm reported for bulk samples, offer better shielding capabilities per unit thickness as well as flexibility in fabrication without recourse to expensive dies. Apart from shields for low magnetic fields, other possible applications include microwave delay lines, magnetic field sensors with moderate sensitivity and liquid nitrogen level sensors.

6. Conclusions

Based on the work reported by us here, one could draw the following conclusions:

i)The as sprayed films are amorphous, even if stoichiometrically pure 1-2-3 powder is used as feed stock. The 1-2-3 SC phase can be readily recovered by appropriate heat treatment of the samples.

ii)The Tc is generally insensitive to the type of spraying carried out and the type of oxide additives added to the YBaCuO powder.

iii)The best Jc results obtainable are limited by the type of spraying technique used. Thus, internal feed, external feed and low pressure plasma spray coatings yielded Jc results of 75 A/cm^2, 200 A/cm^2 and 360 A/cm^2.

iv)The YBaCuO coatings formed by external feed plasma spray are superior to the internal plasma spray in all respects, namely, chemical composition, current density, resistance to flux penetration and stability.

v)Low pressure plasma spray samples exhibit far better Jc values and much better shielding characteristics compared to external feed plasma spray coatings.

vi)Invariably, samples which show better resistance to flux penetration are also more stable when subjected to thermal cycling between $77^{\circ}K$ and RT.

vii)Plasma sprayed coatings of YBaCuO can be readily used for shielding low magnetic fields.

ACKNOWLEDGEMENT

We wish to thank the BHEL management for their support and permission. We also thank Department of Science and Technology for their financial support. Thanks are due to Dr.V.Ramamurthy for his suggestions.

REFERENCES

1. M.K.Wu, J.R.Ashburn, C.J.Torng, P.H.Hor, R.L.Meng, L.Gao, Z.J.Huang, Y.Q.Wang and C.W.Chu, *Phys. Rev. Lett.* **58**(1987)908.
2. R.W.McCallum, J.D.Verhoven, M.A.Noack, E.D.Gibson, F.C.Laabs, D.K.Finnemore and A.R.Moodenbaugh, *Adv. Ceram. Mater.*2(1987)388.
3. Kohno, Y.Ikeno, N.Sadakata and K.Goto, *Jpn. J. Appl. Phys. Lett.* **27** (1988)L77.
4. H.Sekine, K.Inoue, H.Maeda, K.Numata, K.Mori and H.Yamamoto, *Appl. Phys. Lett.* **52** (1988)2261.
5. J.P.Kirkland, R.A.Neiser, H.Herman, W.T.Elam, S.Sampath, E.F.Skelton, D.Gansert and H.G.Wang, *Adv. Ceram. Mater.* **2**(1987)401.
6. J.J.Cuomo, C.R.Guarnieri, S.A.Shivashankar, R.A.Roy, D.S.Yee and R.Rosenberg, *Adv. Ceram. Mater.* **2**(1987)401.
7. Tackikawa, I.Wattanabe, S.Kosuge, M.Kabasawa, T.Suzuki, Y.Matsuda, and Y.Shinbo, *Appl. Phys. Lett.* **52**(1988)1011.
8. T.Konaka, I.Sankawa, T.Matsuura, T.Higashi and K.Ishihara, *Jpn. J. Appl. Phys. Lett.* **27**(1988)L1092.
9. R Suryanarayanan, G Brun, M Bouanani, F Meguellat, M Rateau, H Pankowska, O Gorochovand G T Bhandage, Proc. of the 12th International Conf. on Thermal Spraying, paper 47, June 1989, London, UK.
10. K.Sato, T.Hikata, H.Mukal, M.Ueyama, N.Shibuta, T.Kato, T.Masuda, M.Nagata, K.Iwata and T.Mitsui, *IEEE Trans. on Mag., Vol.* **27**, *No. 2*, March 1991
11. D.Maheswar Reddy, G.Swaminathan, S.Rajendra Kumar, K.Venugopal, S.Ravichandran, M.V.T.Dhananjeyan and S.C.Mohan, *Thin Solid Films*, **206**(1991)169-174
12. K.A.Durgaprasad, K.C.Lahiry, D.Maheswar Reddy, S.C.Mohan, S.Rajendra Kumar, N.Ramadas, S.Ravichandran, R.Somasundaram, G.Swaminathan and K.Venugopal, *Bull. Mater. Sci.* **14**(1990)12

SUBJECT INDEX